This book belongs to

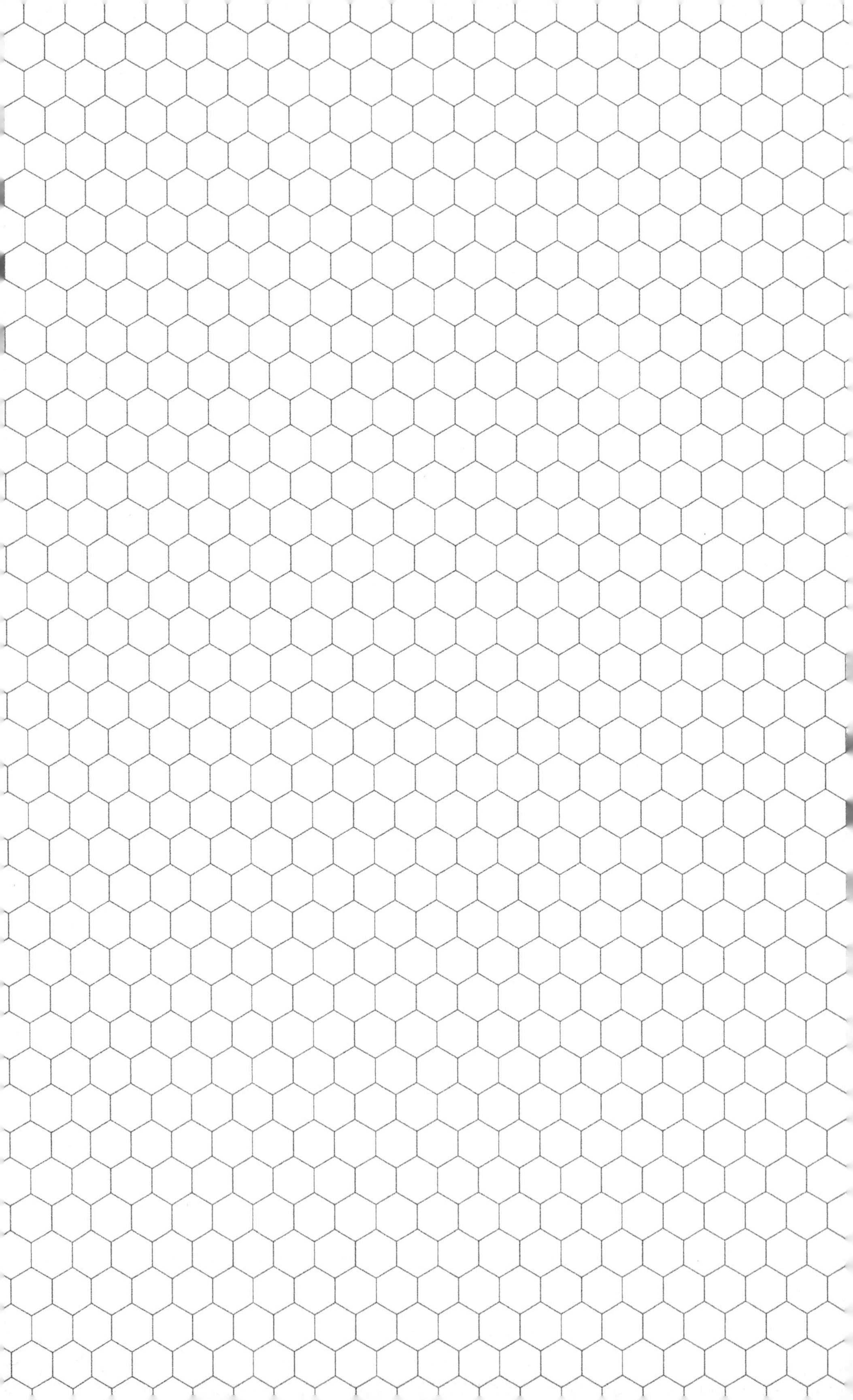

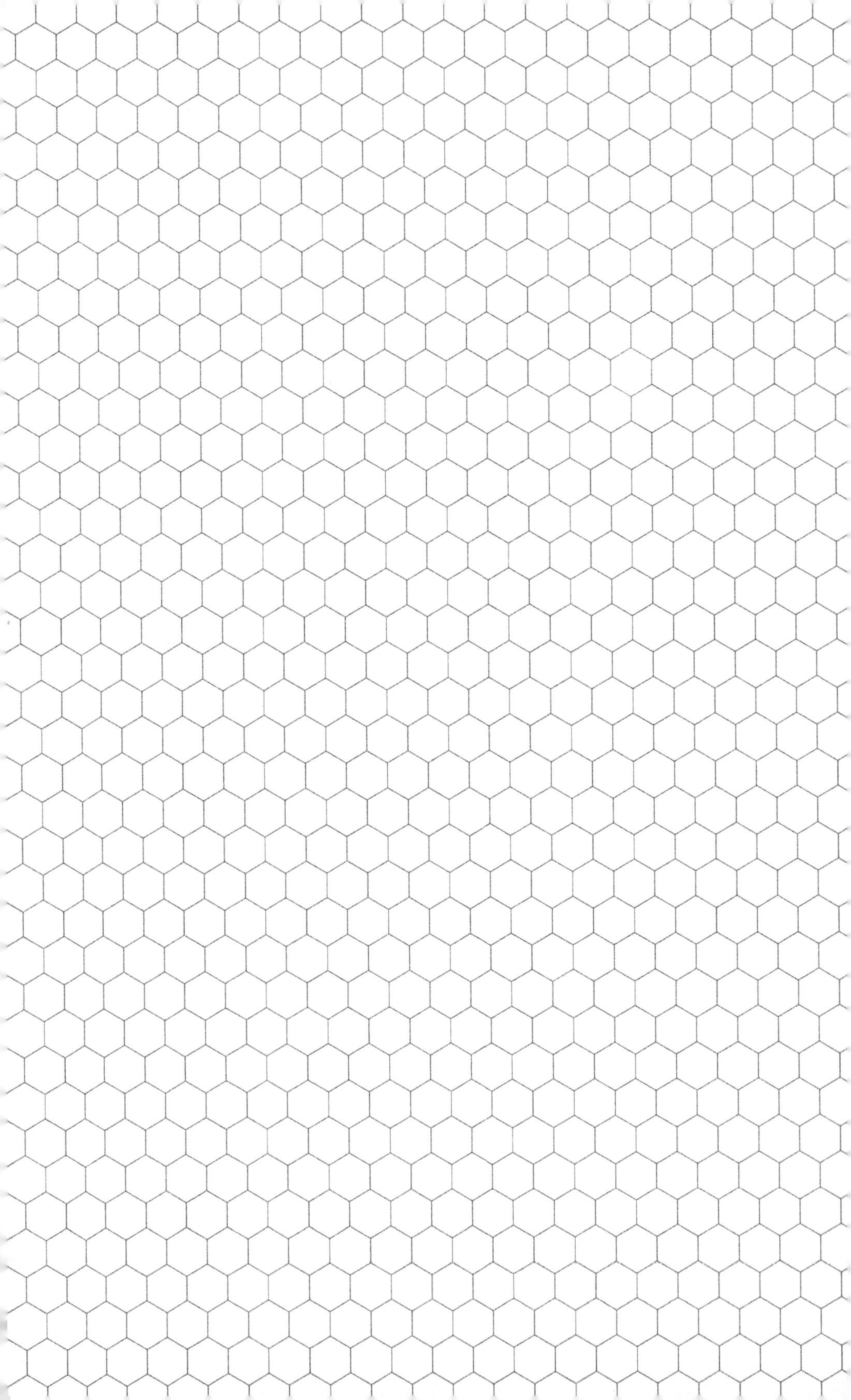

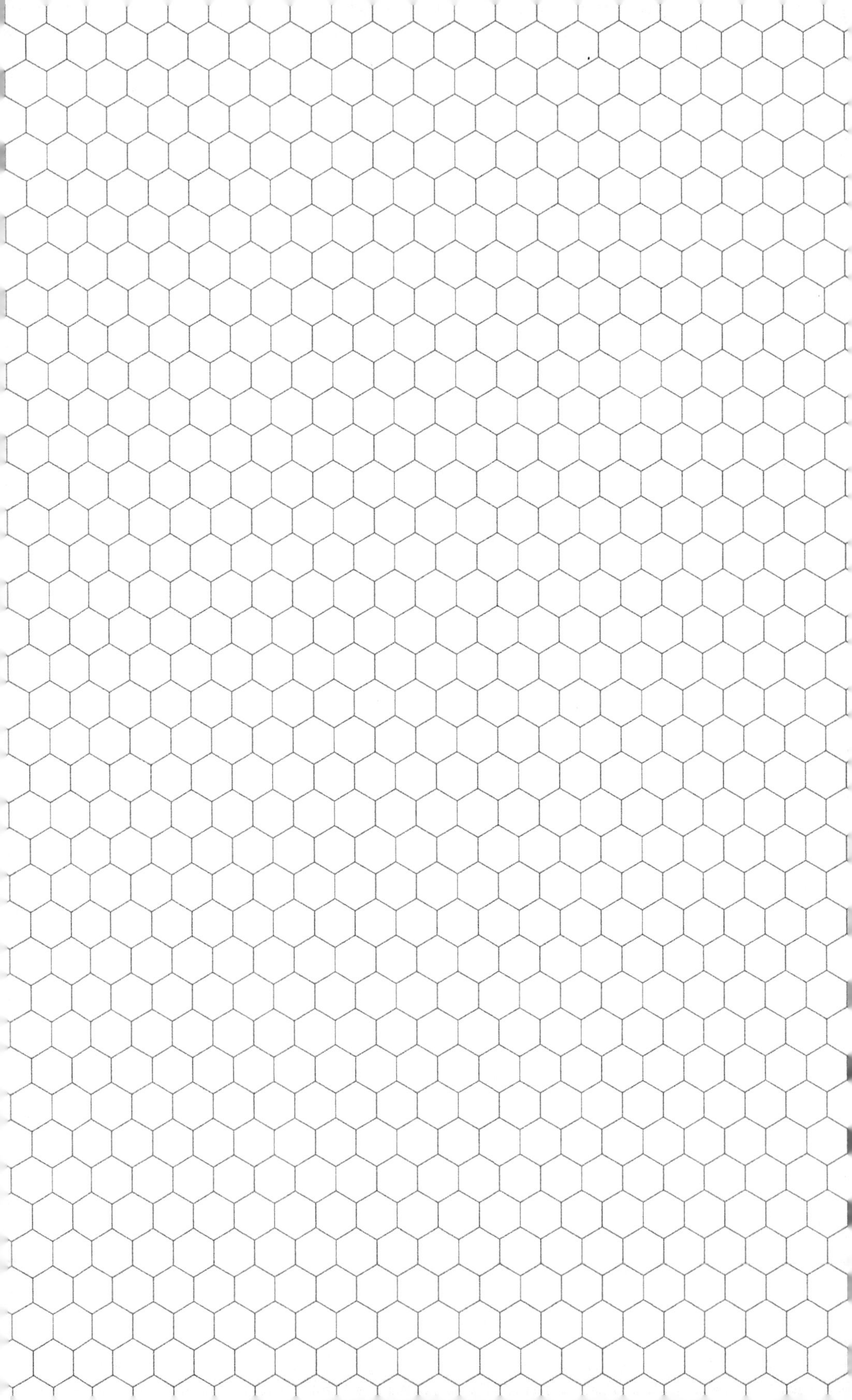

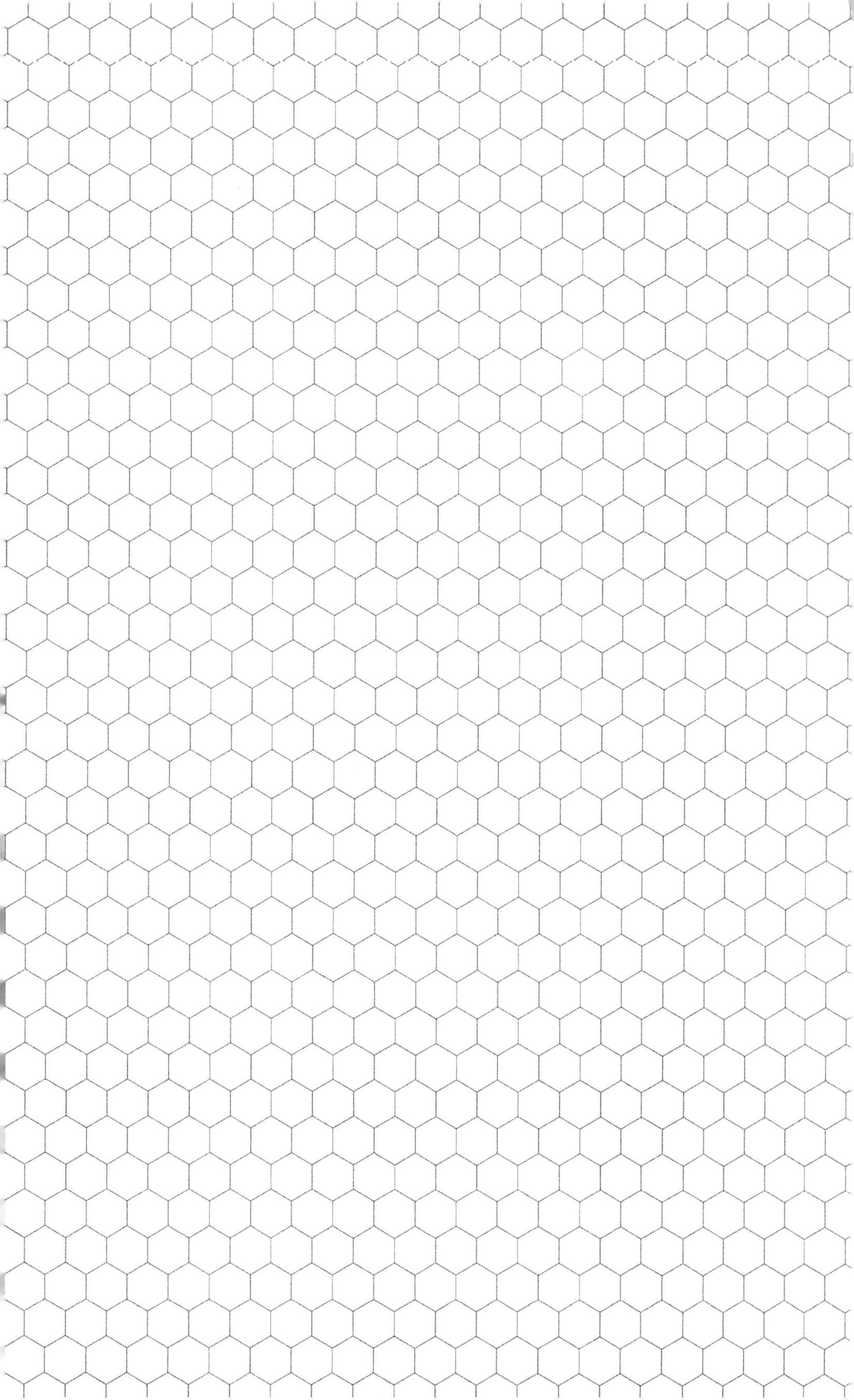

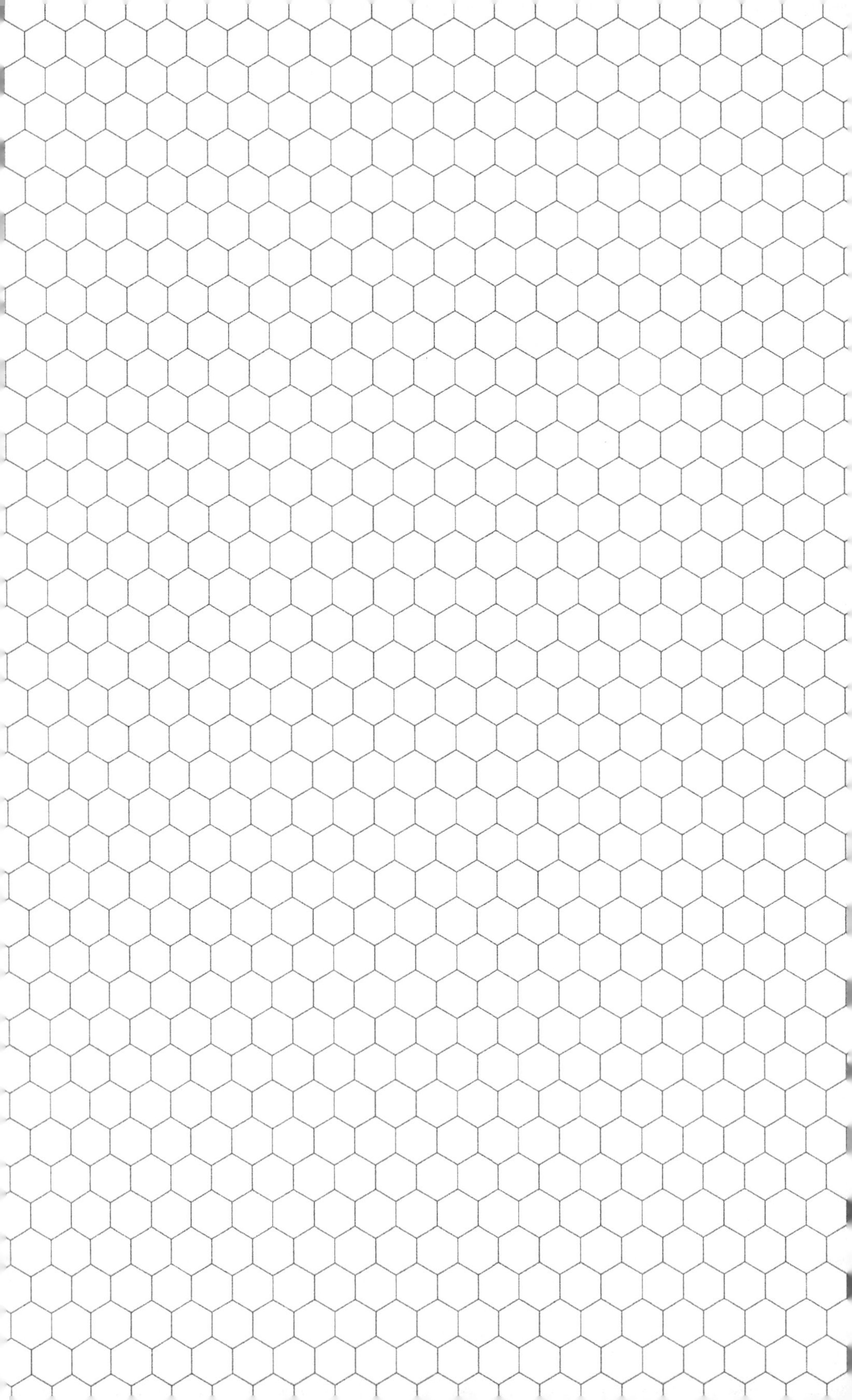

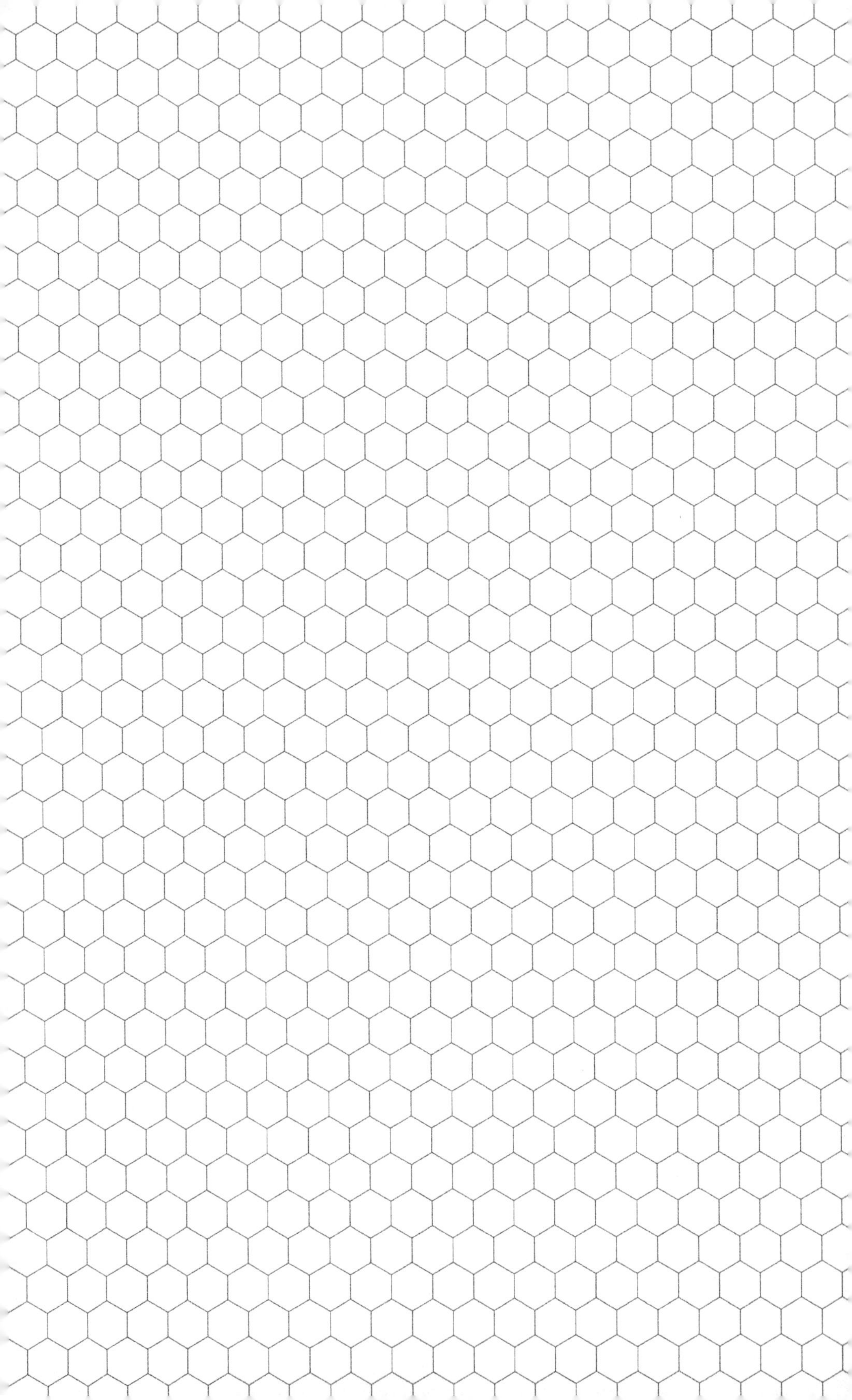

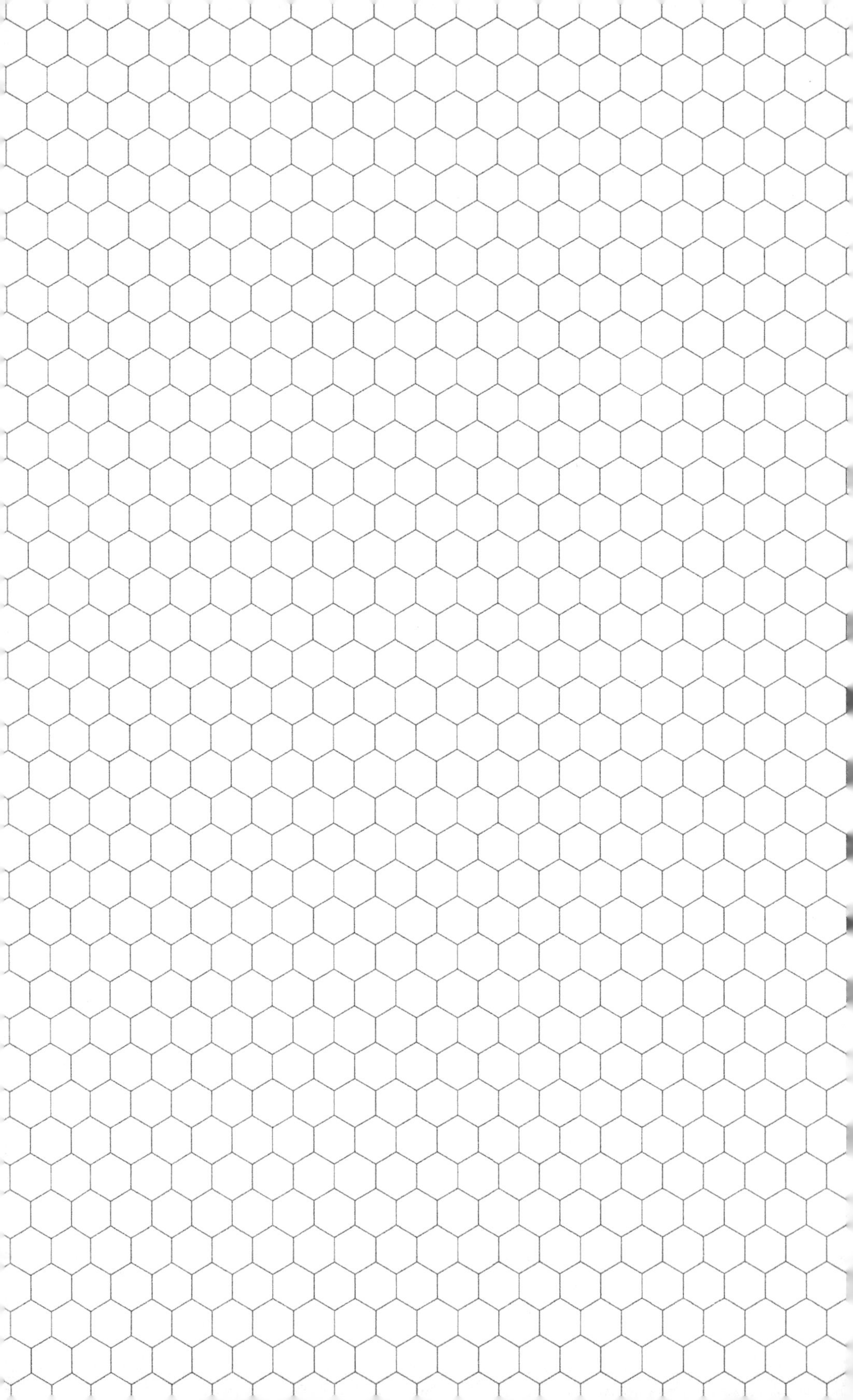

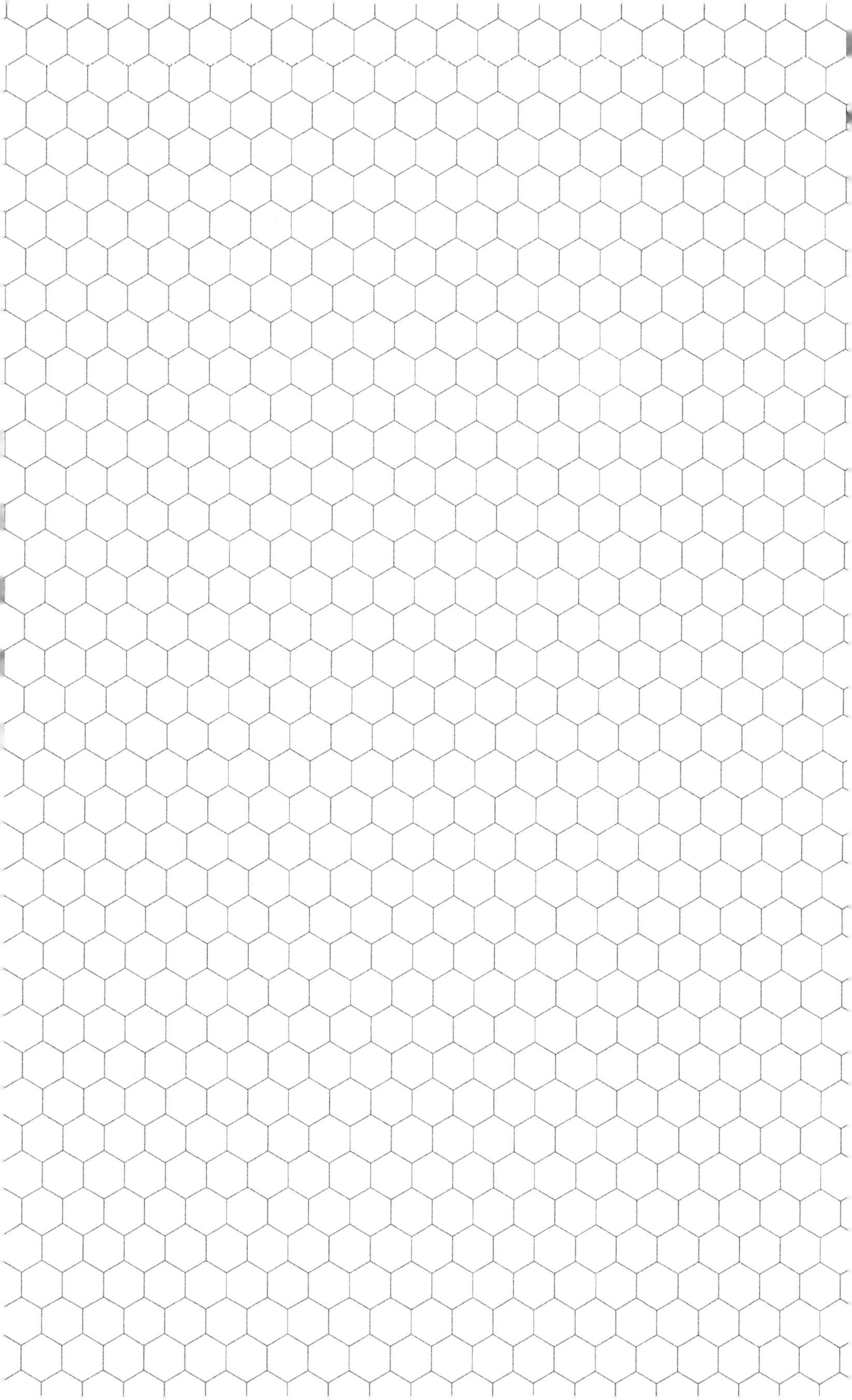

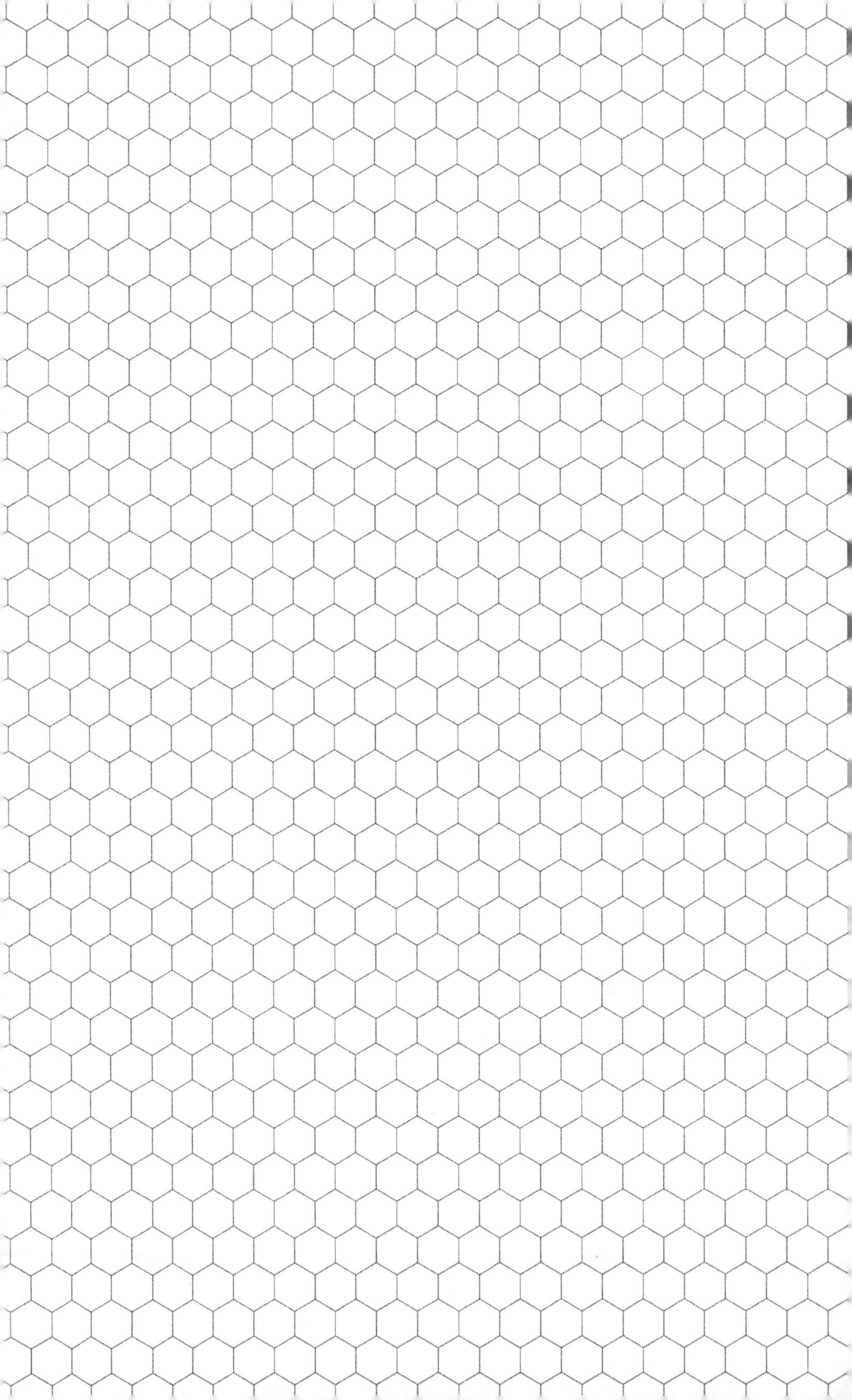

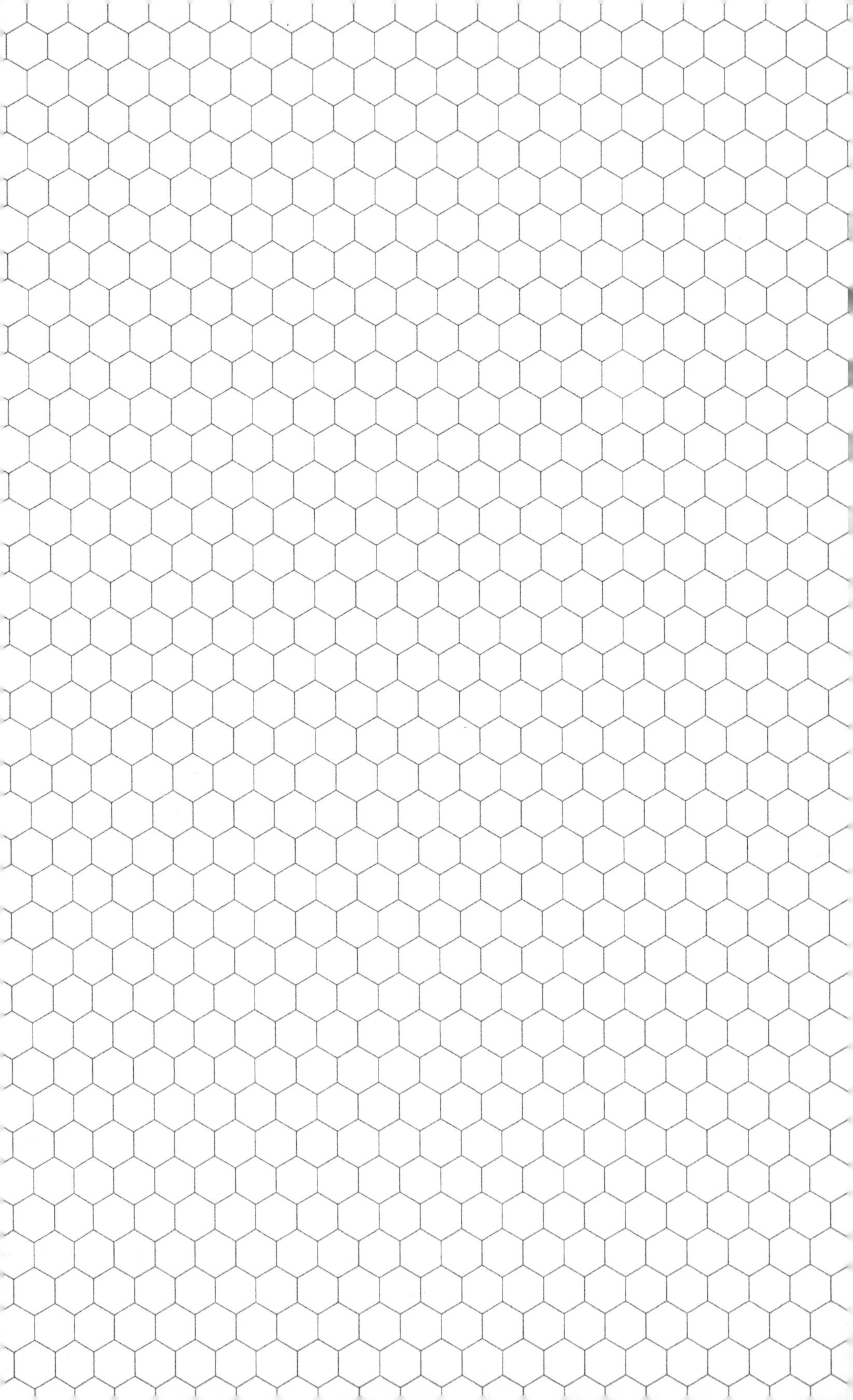

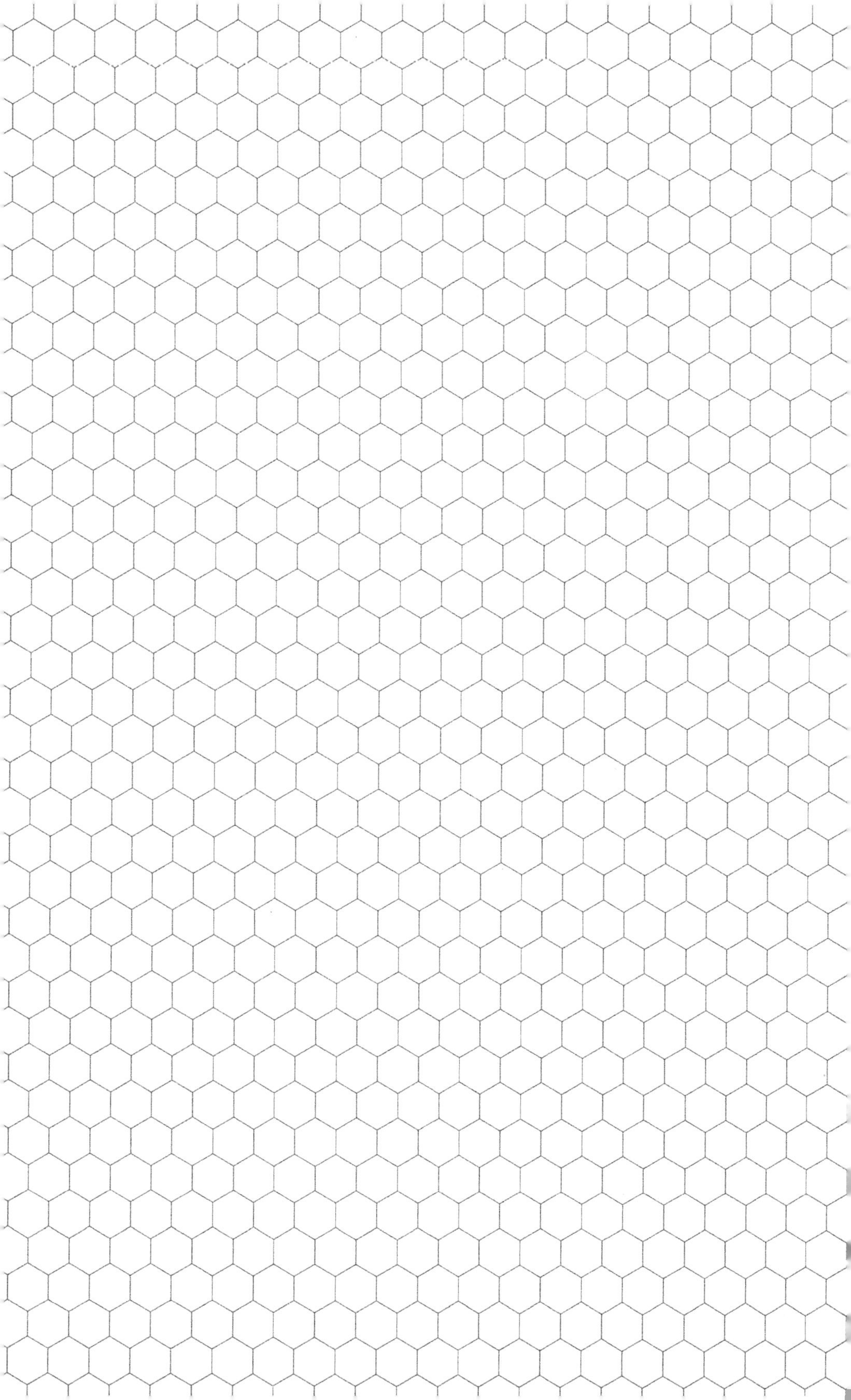

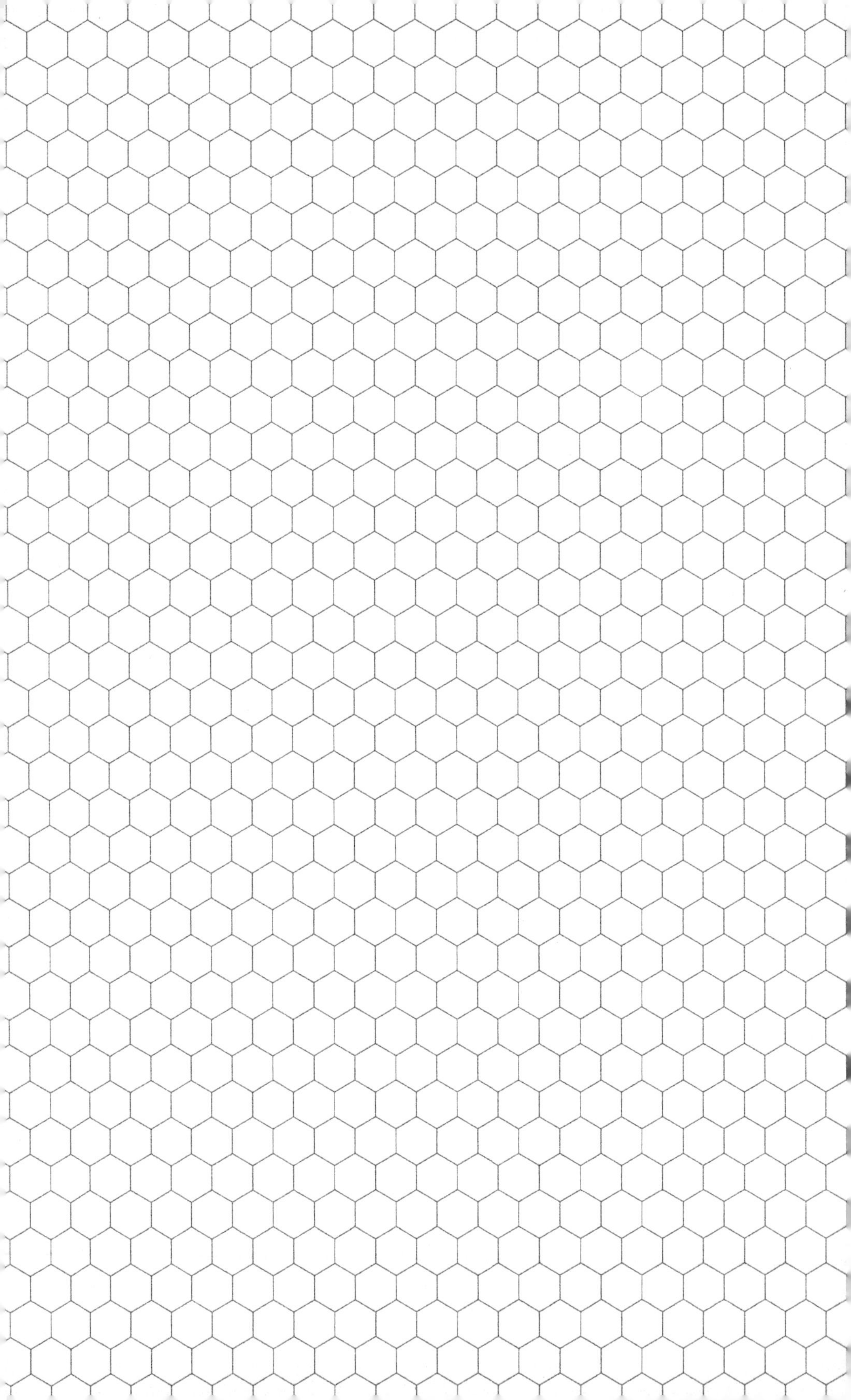

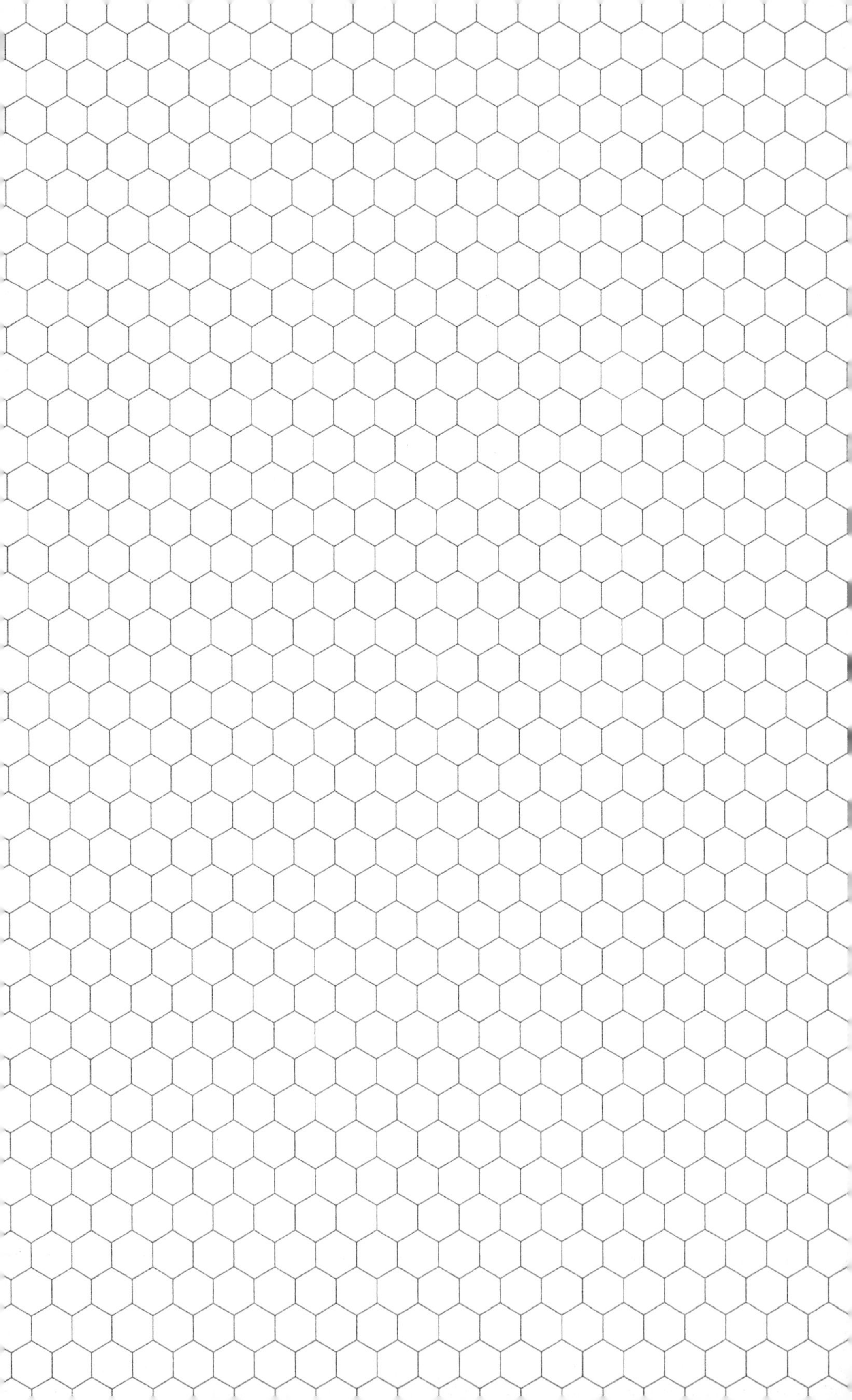

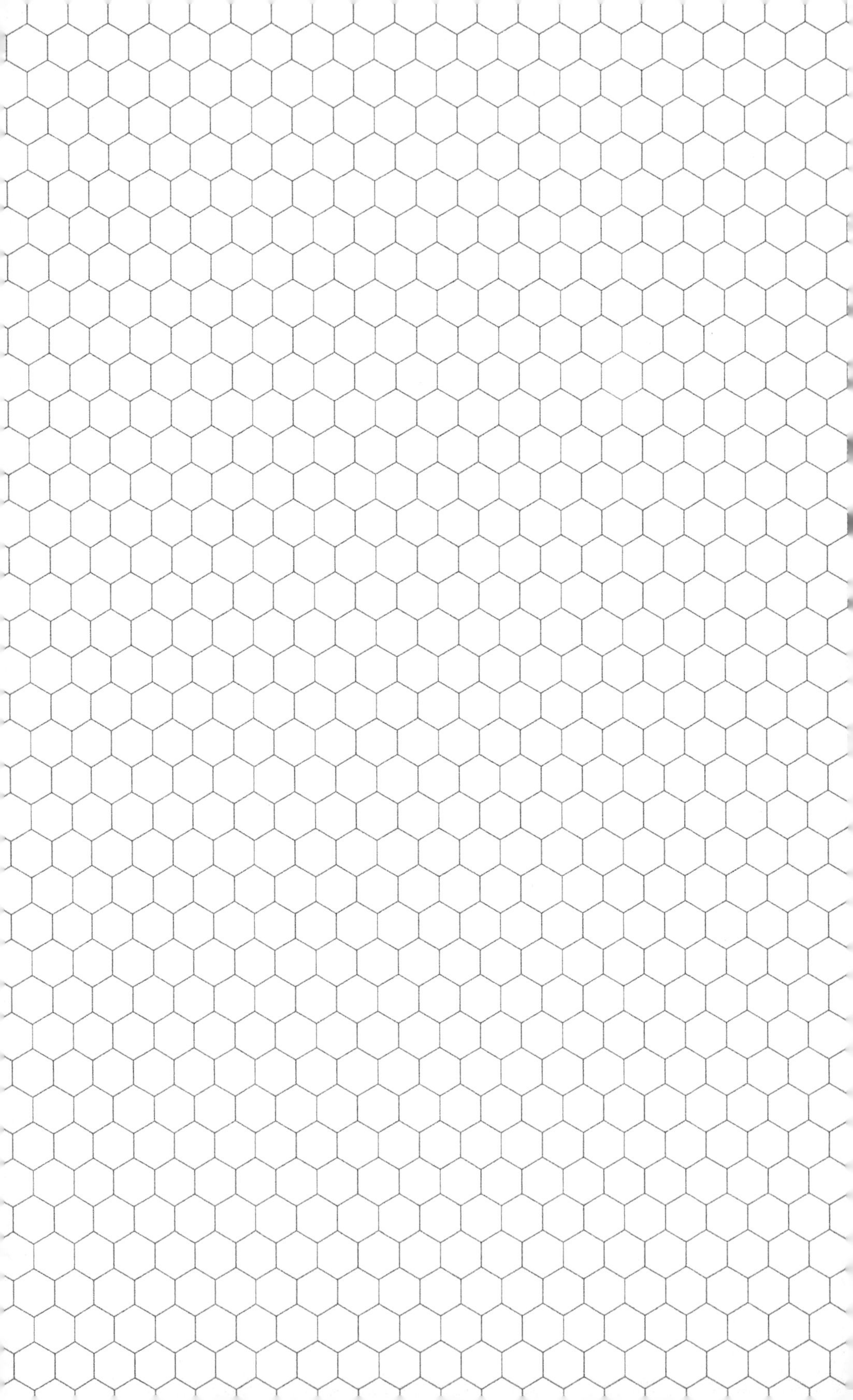

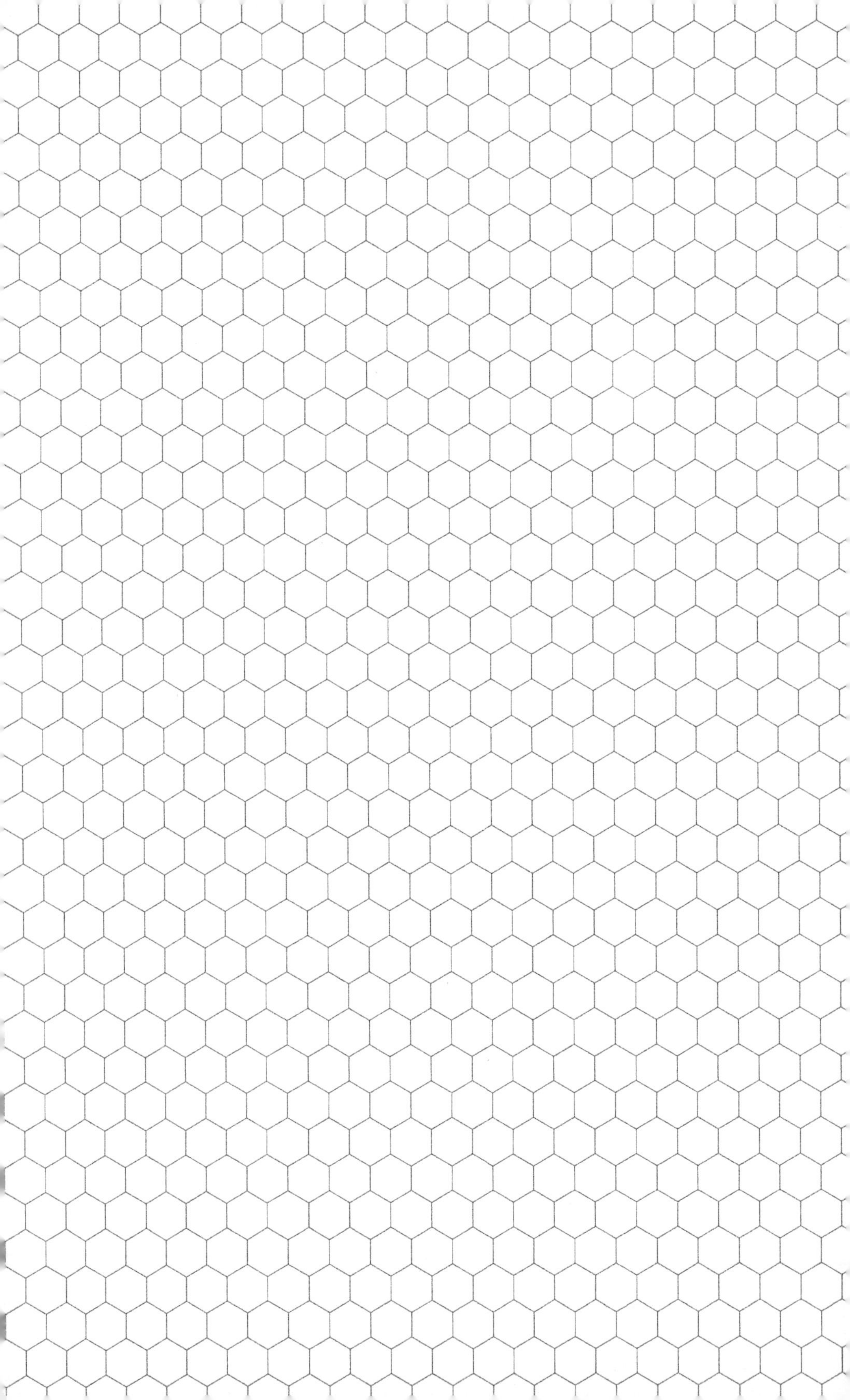

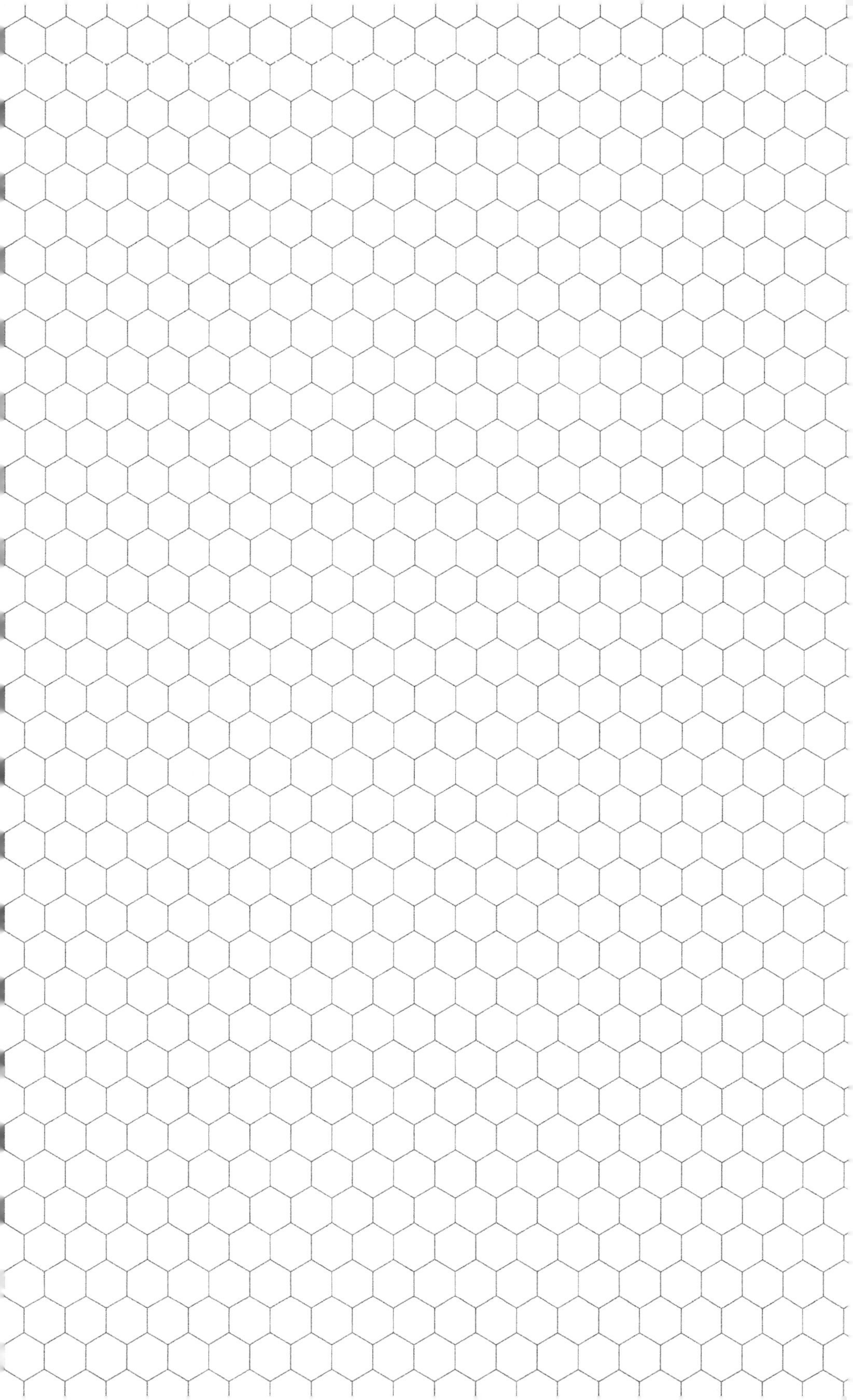

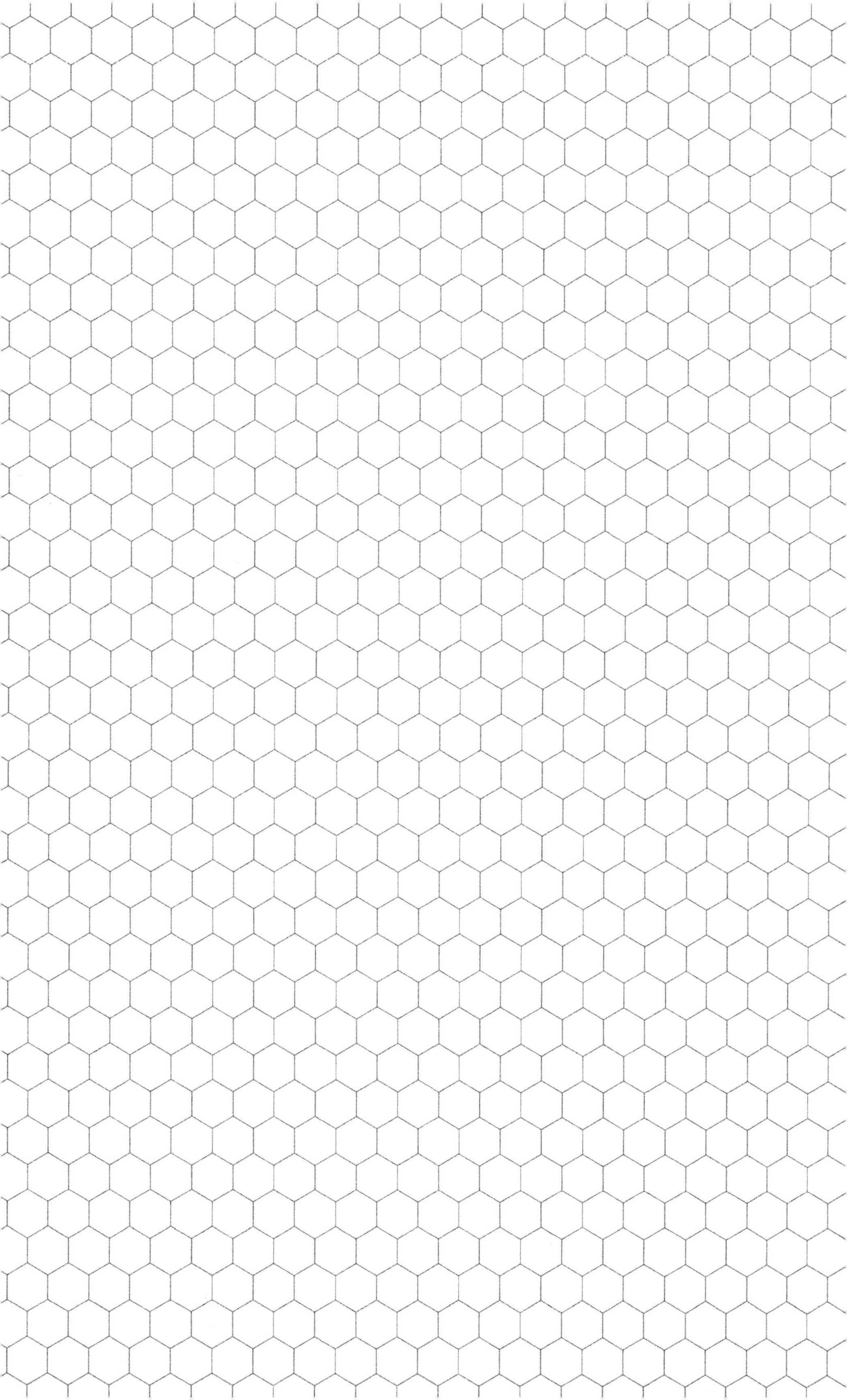

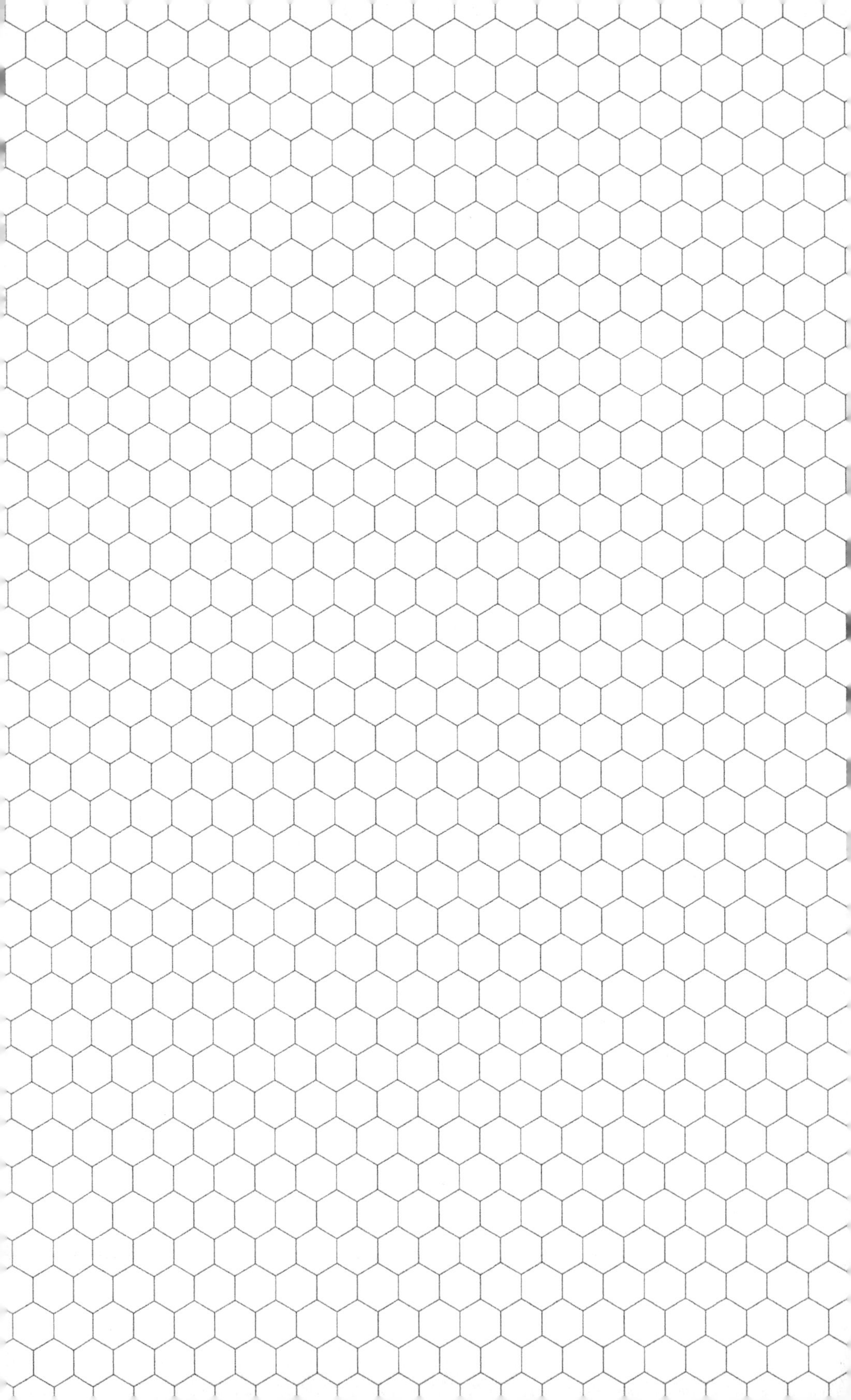

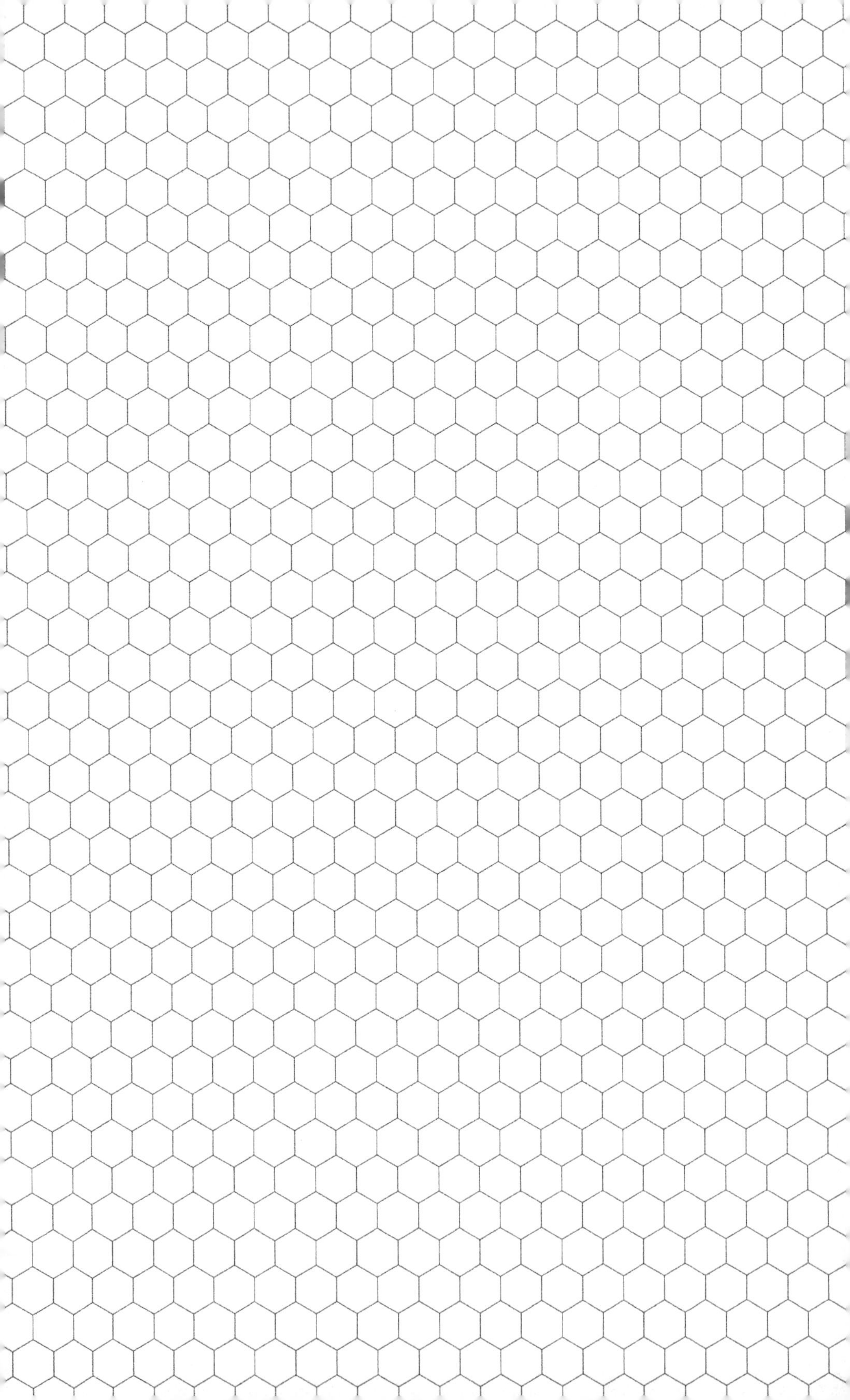

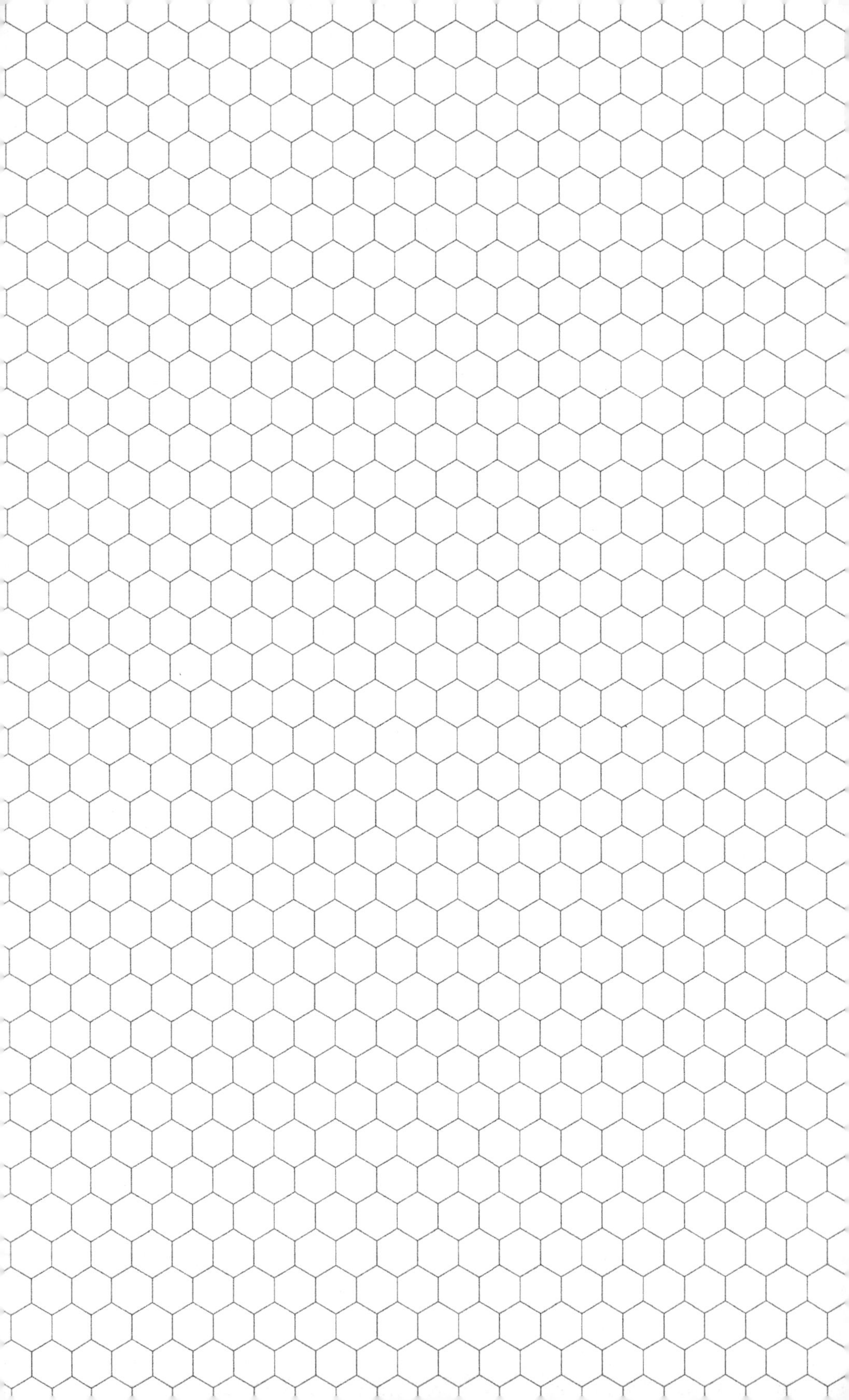

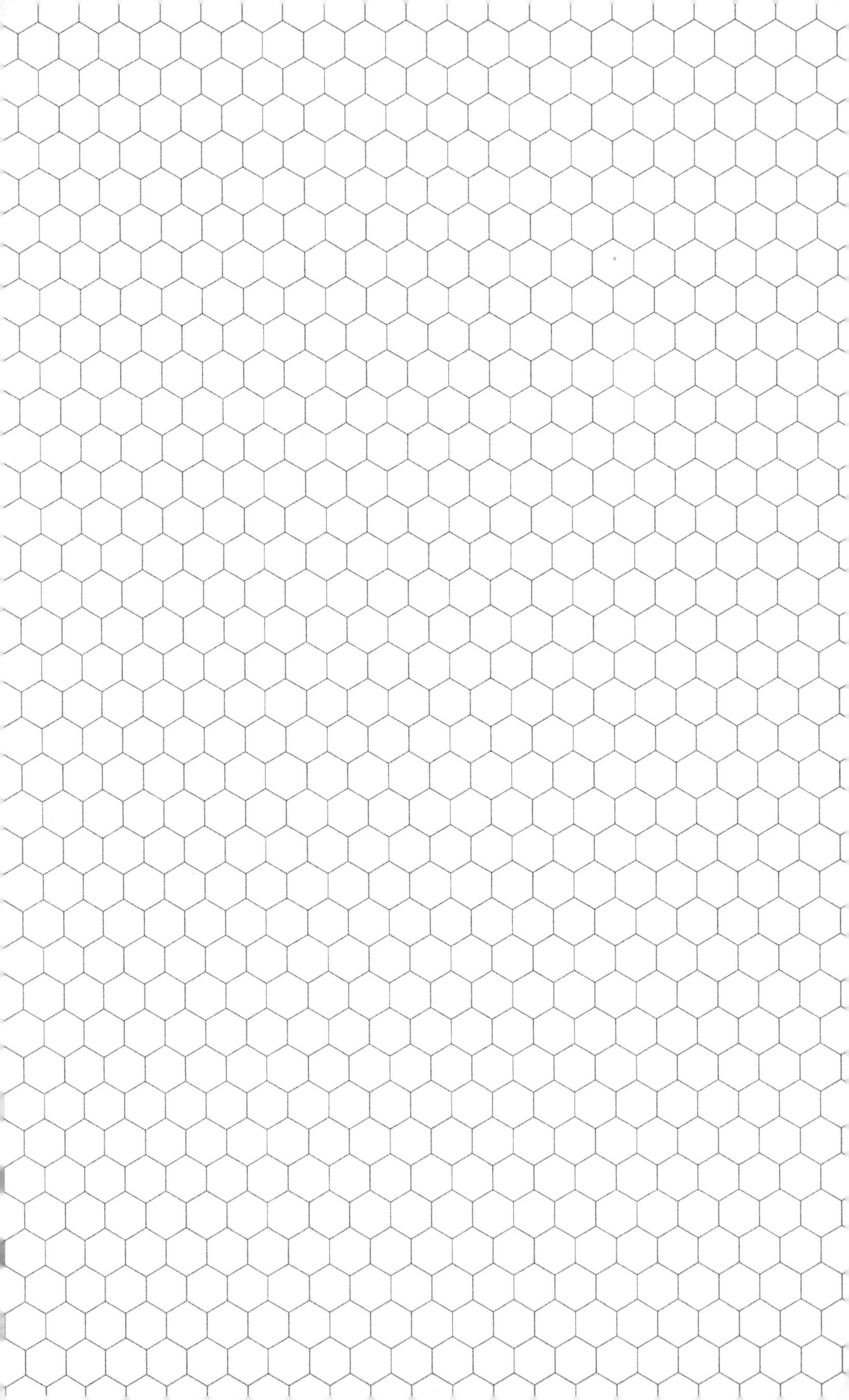

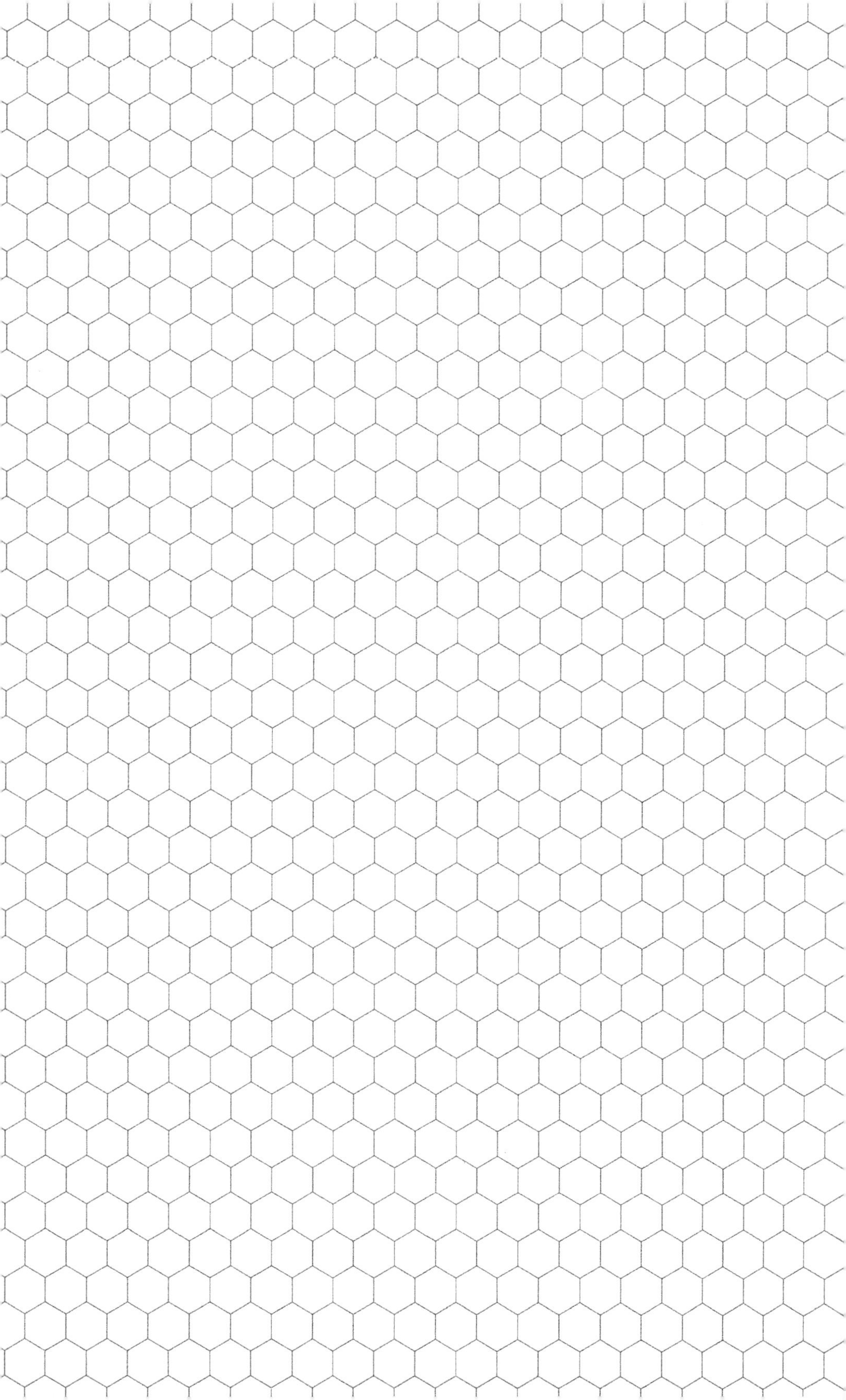

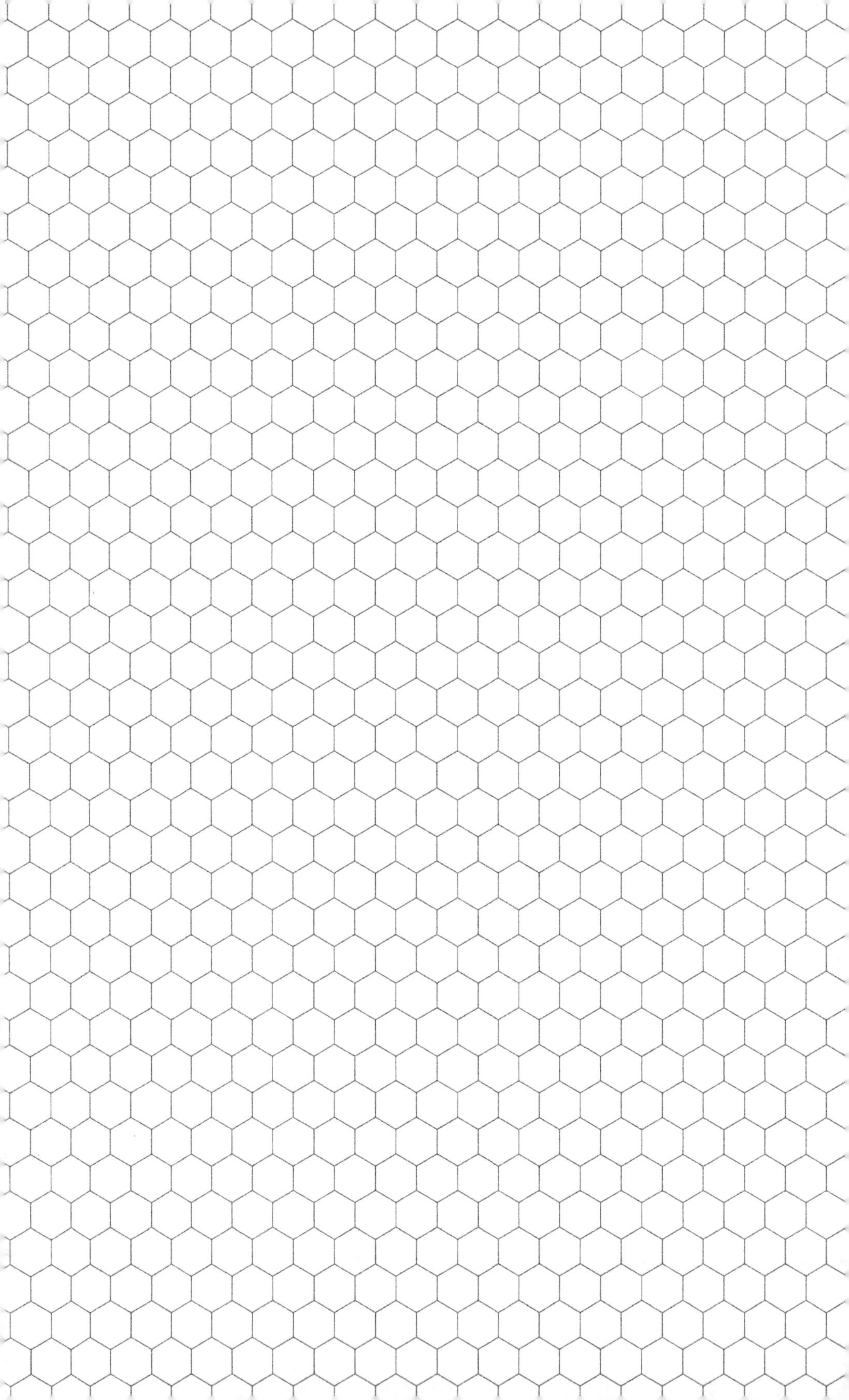

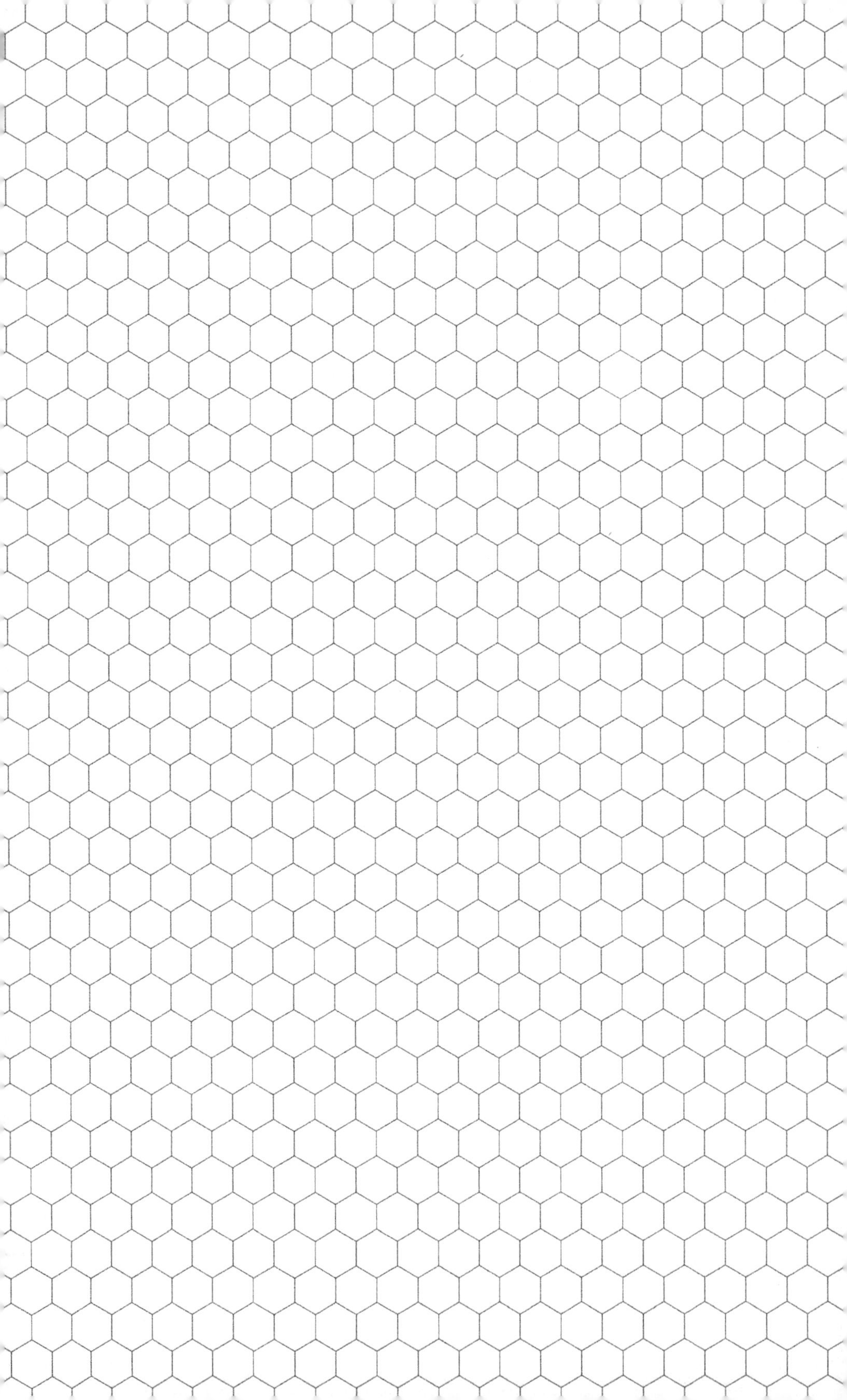

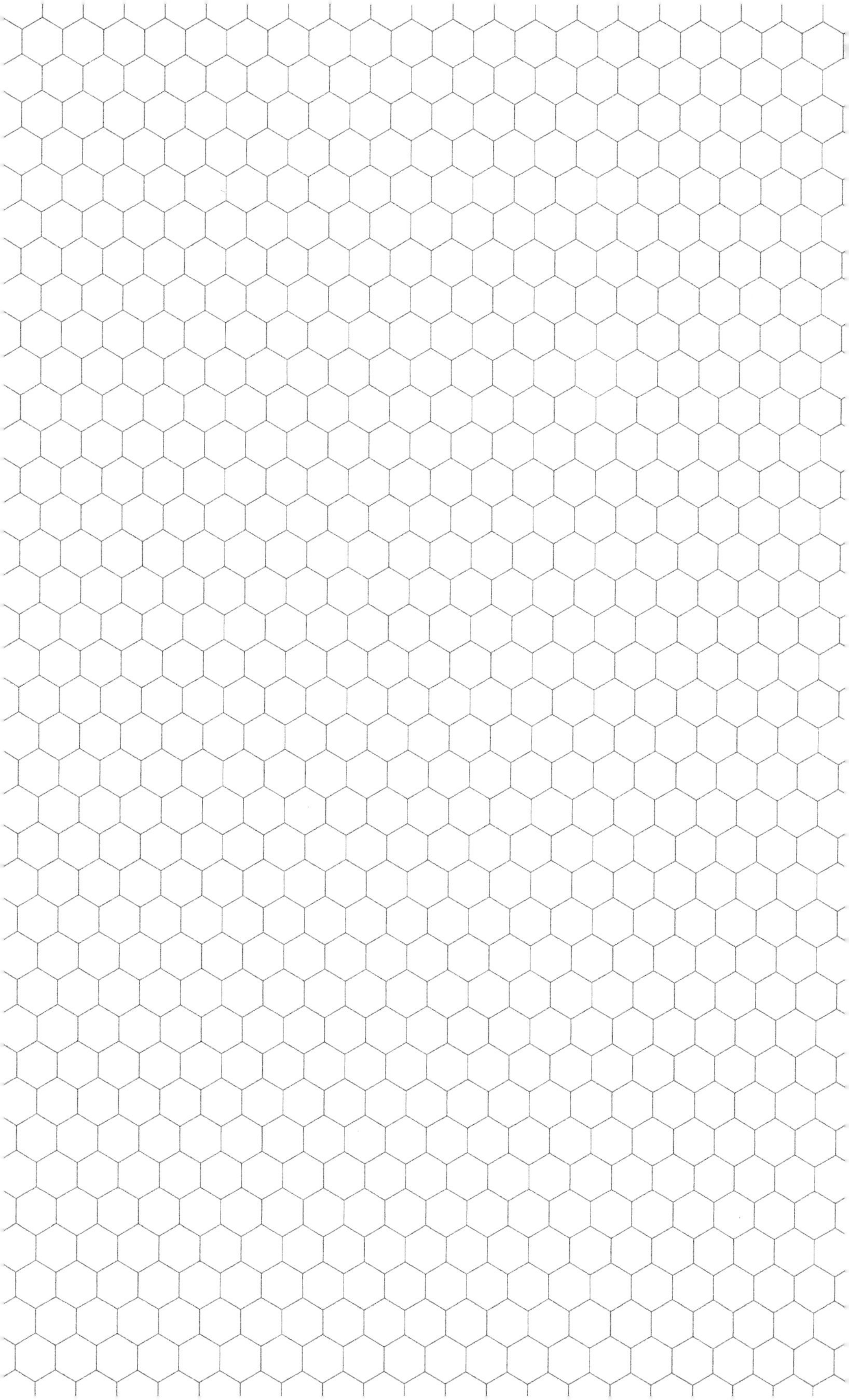

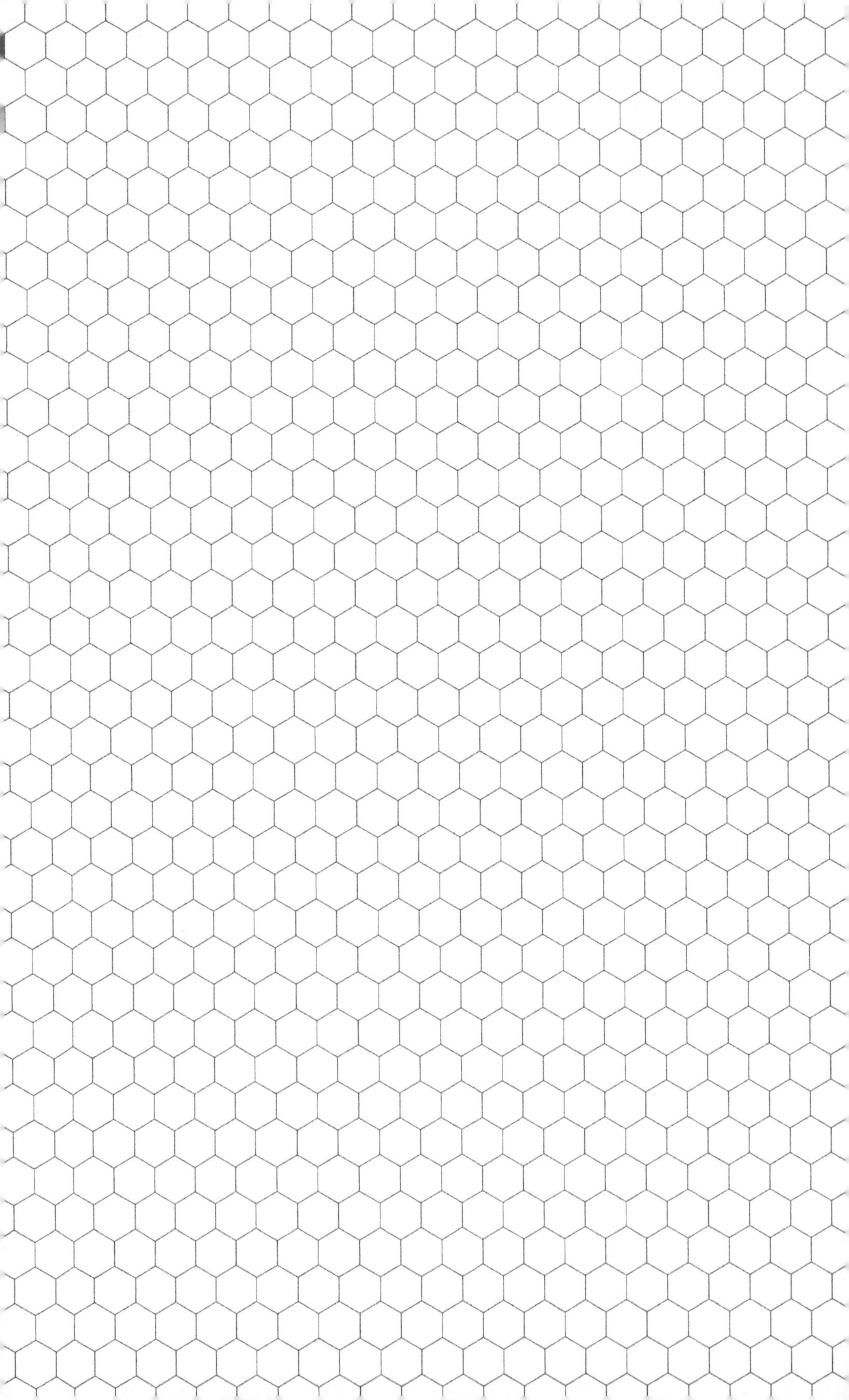

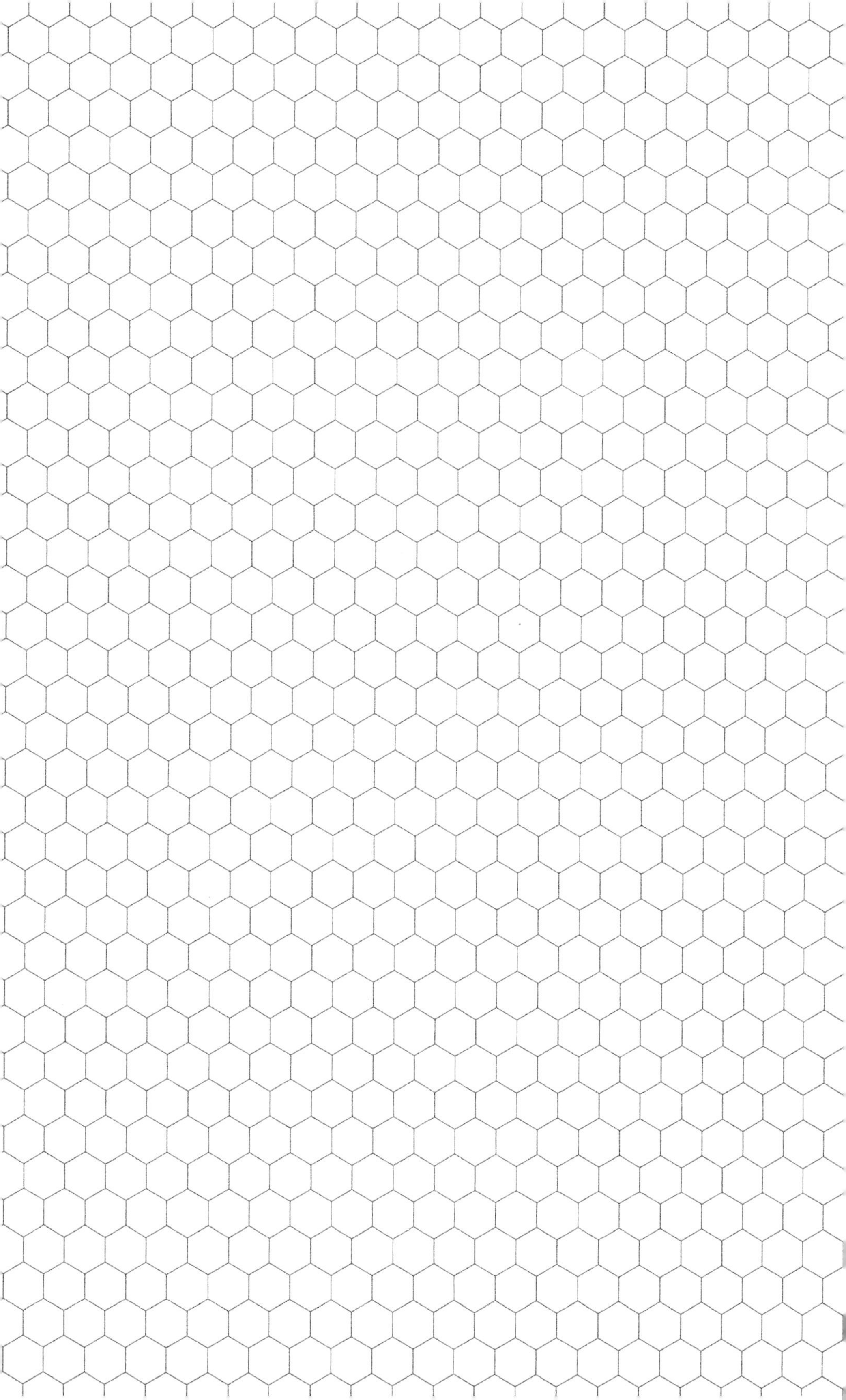

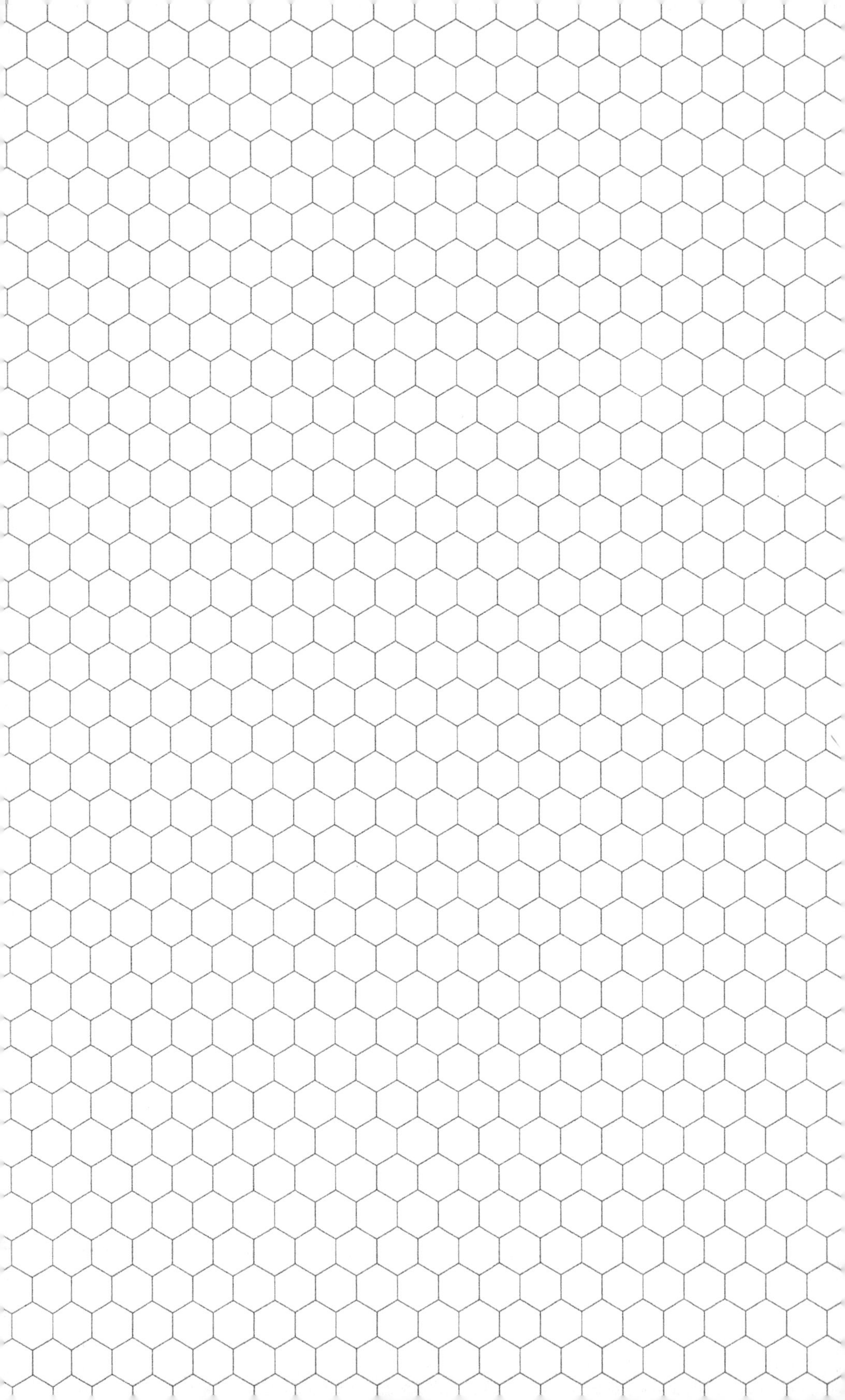

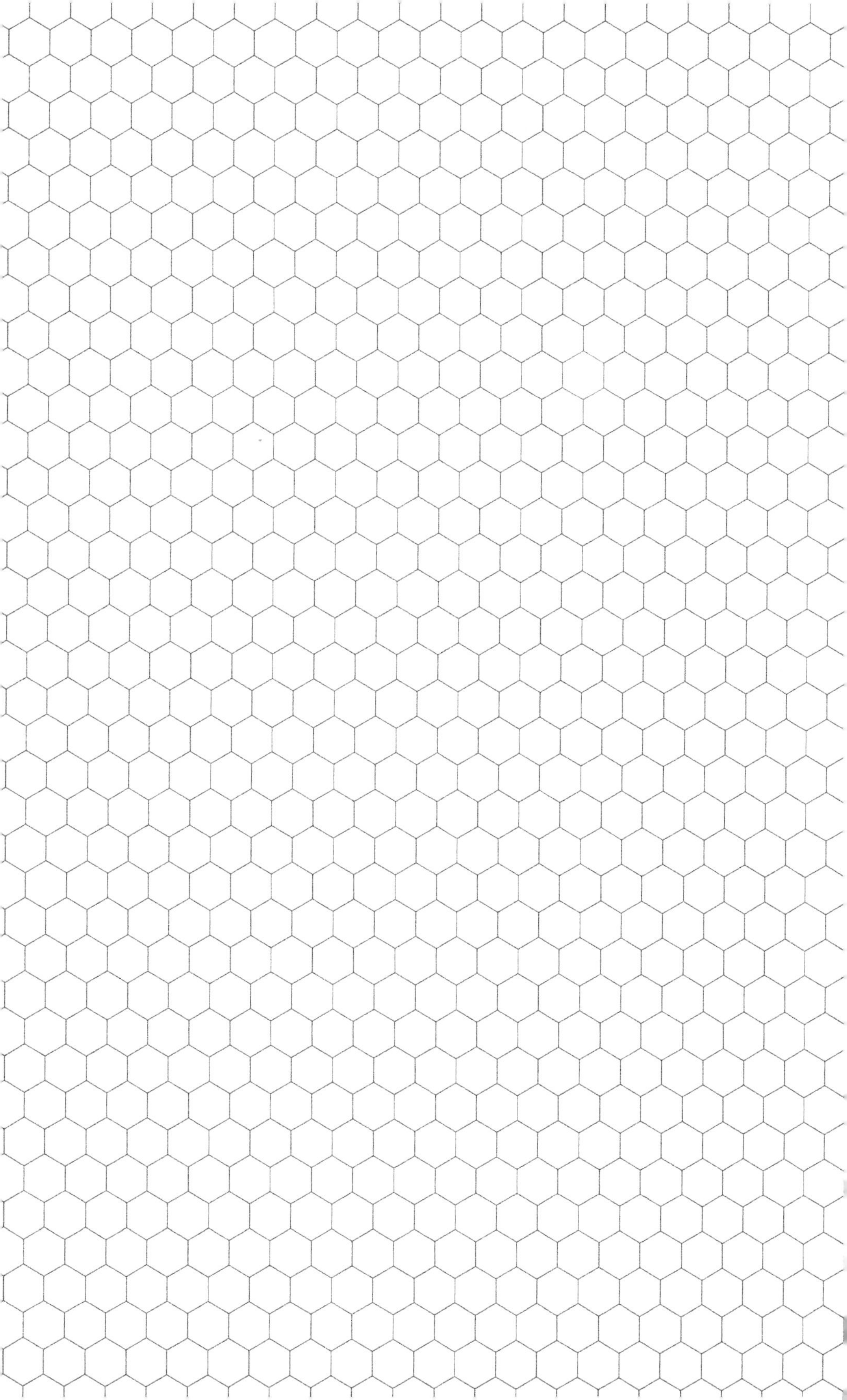

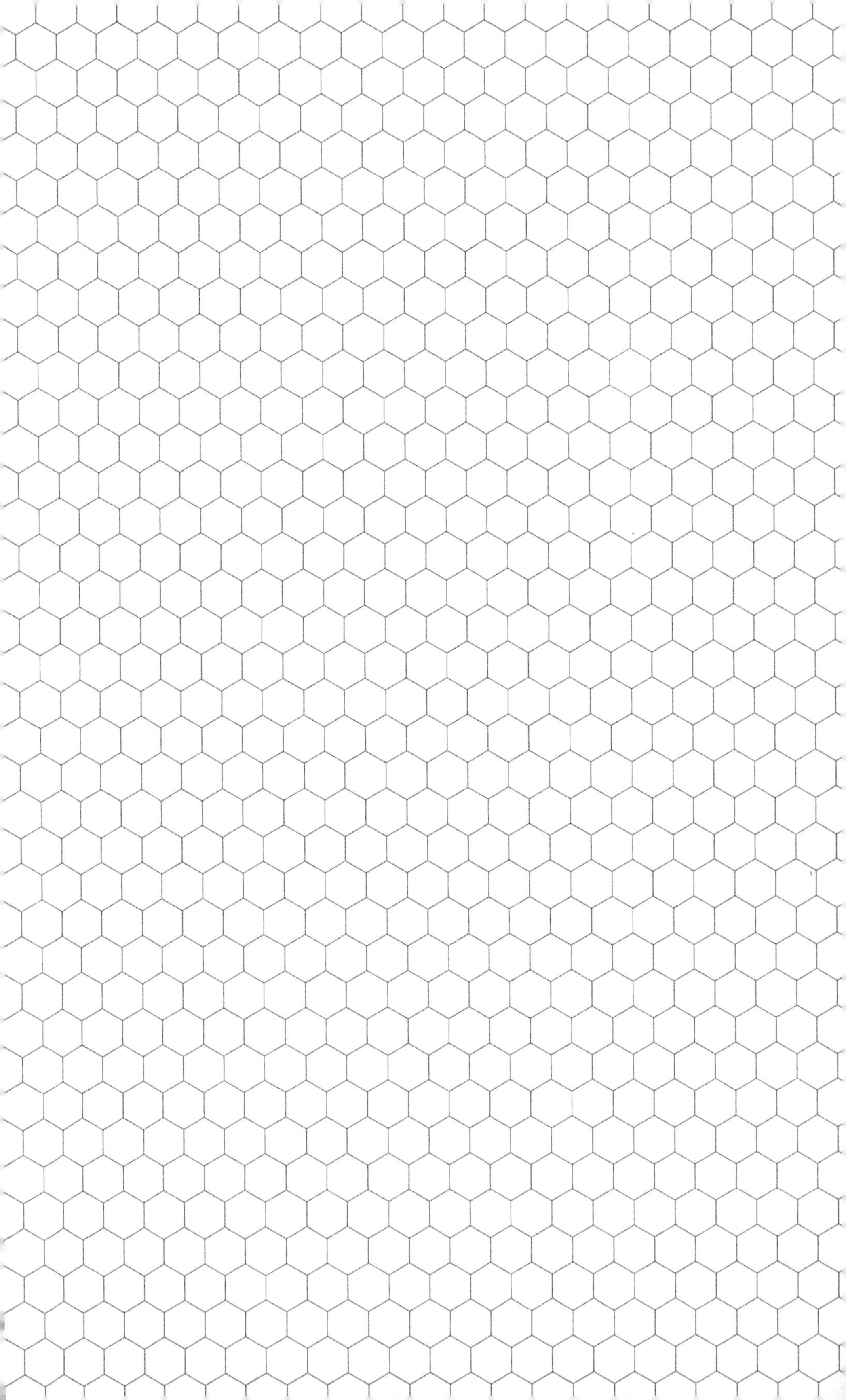

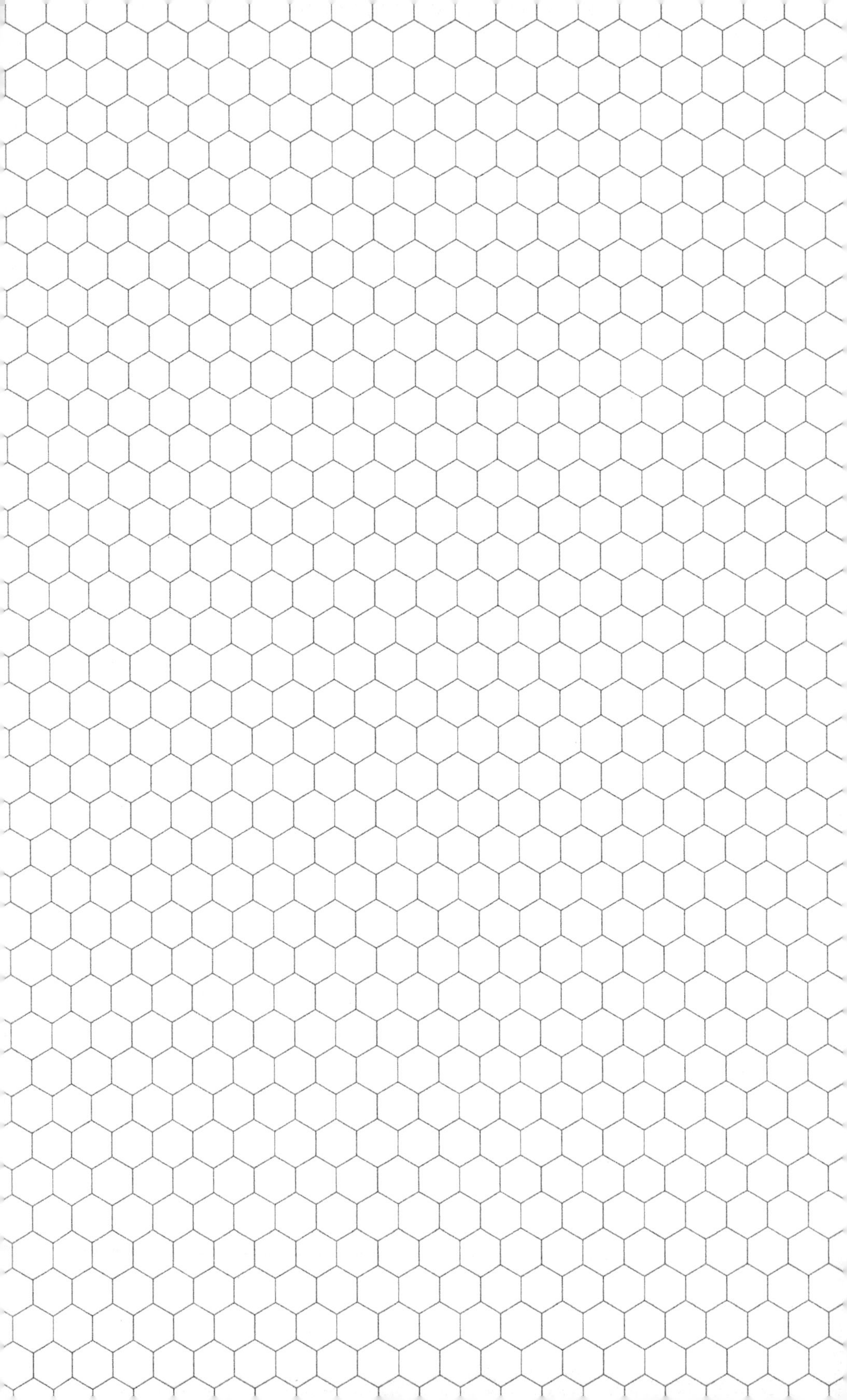

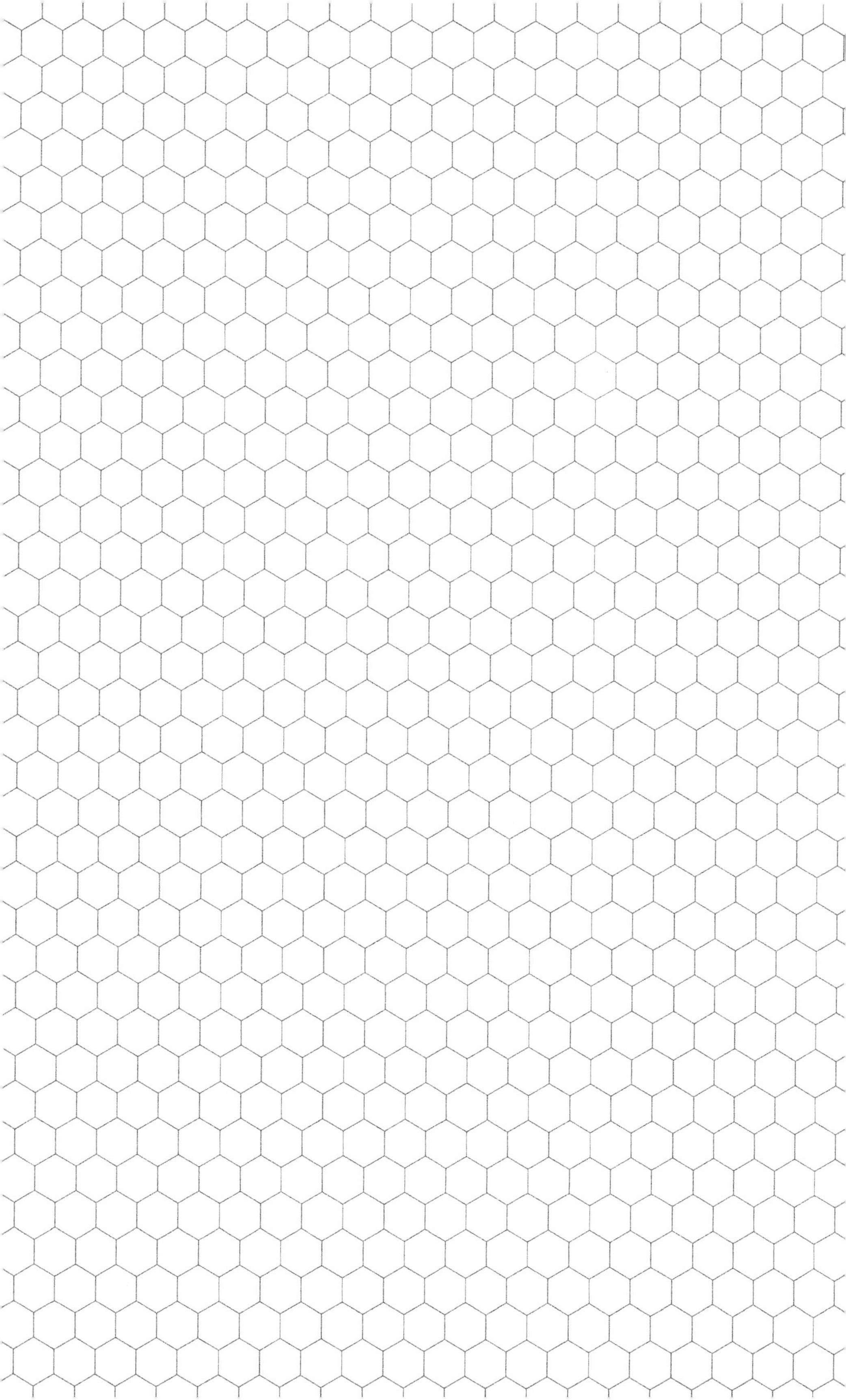

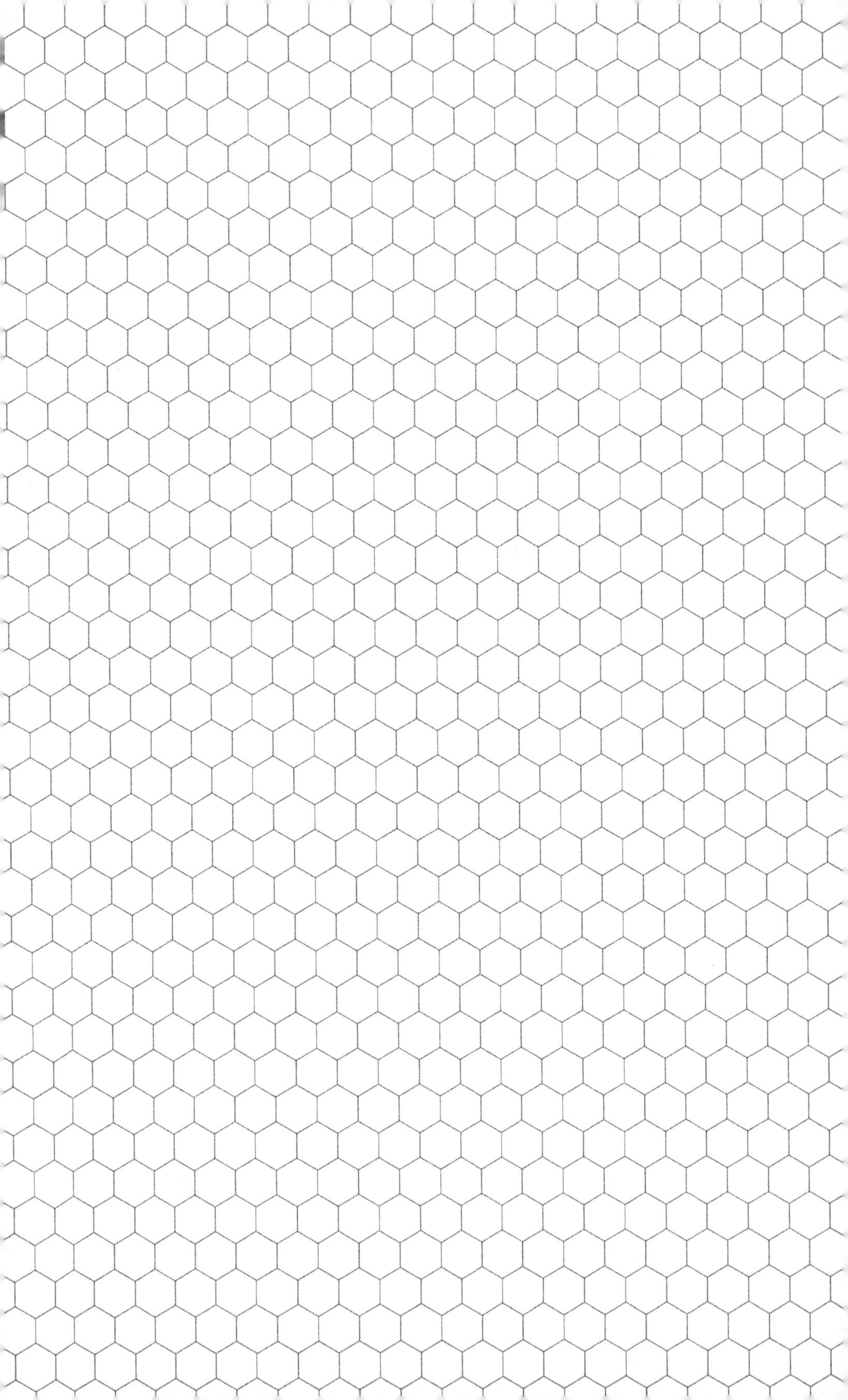

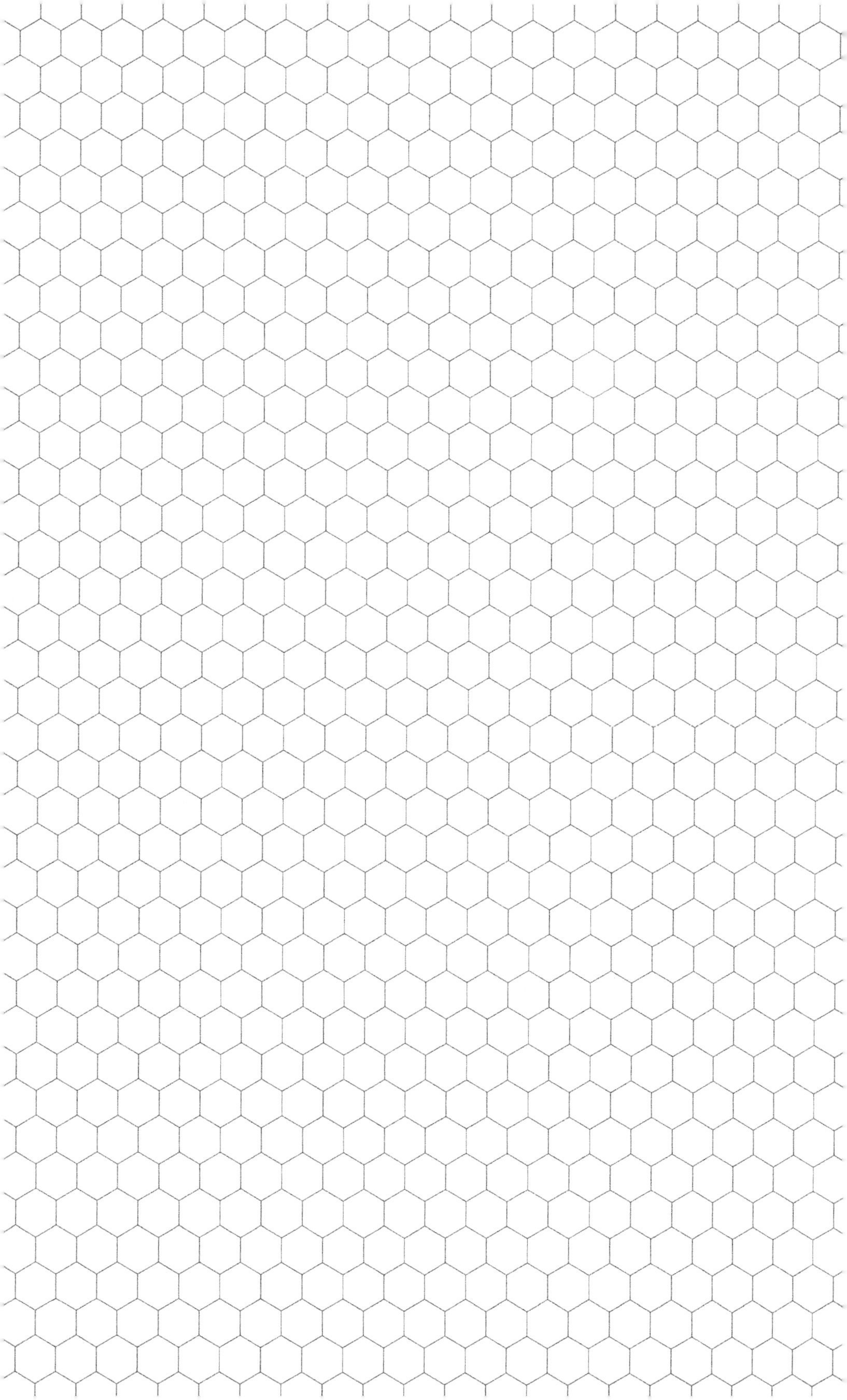

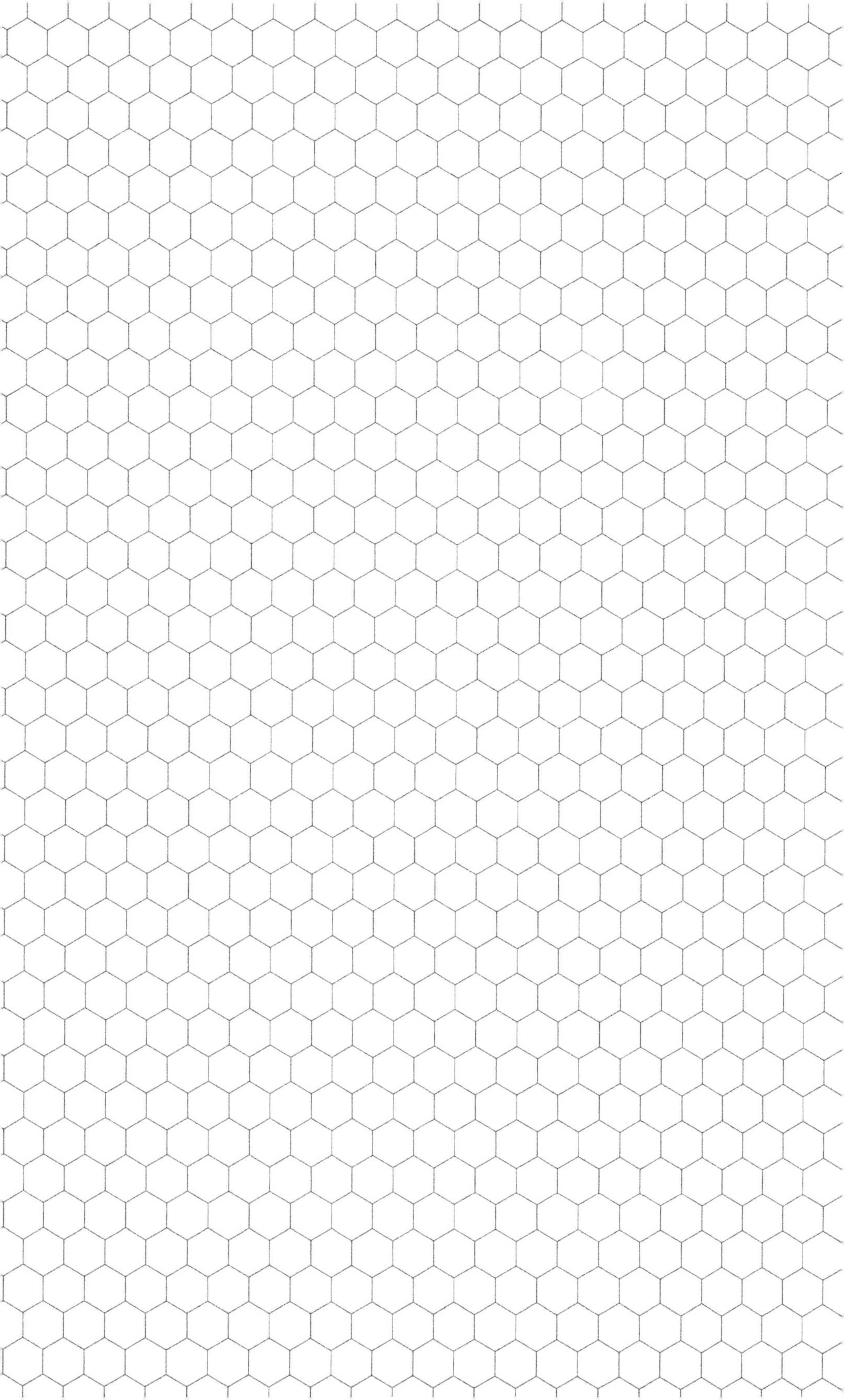

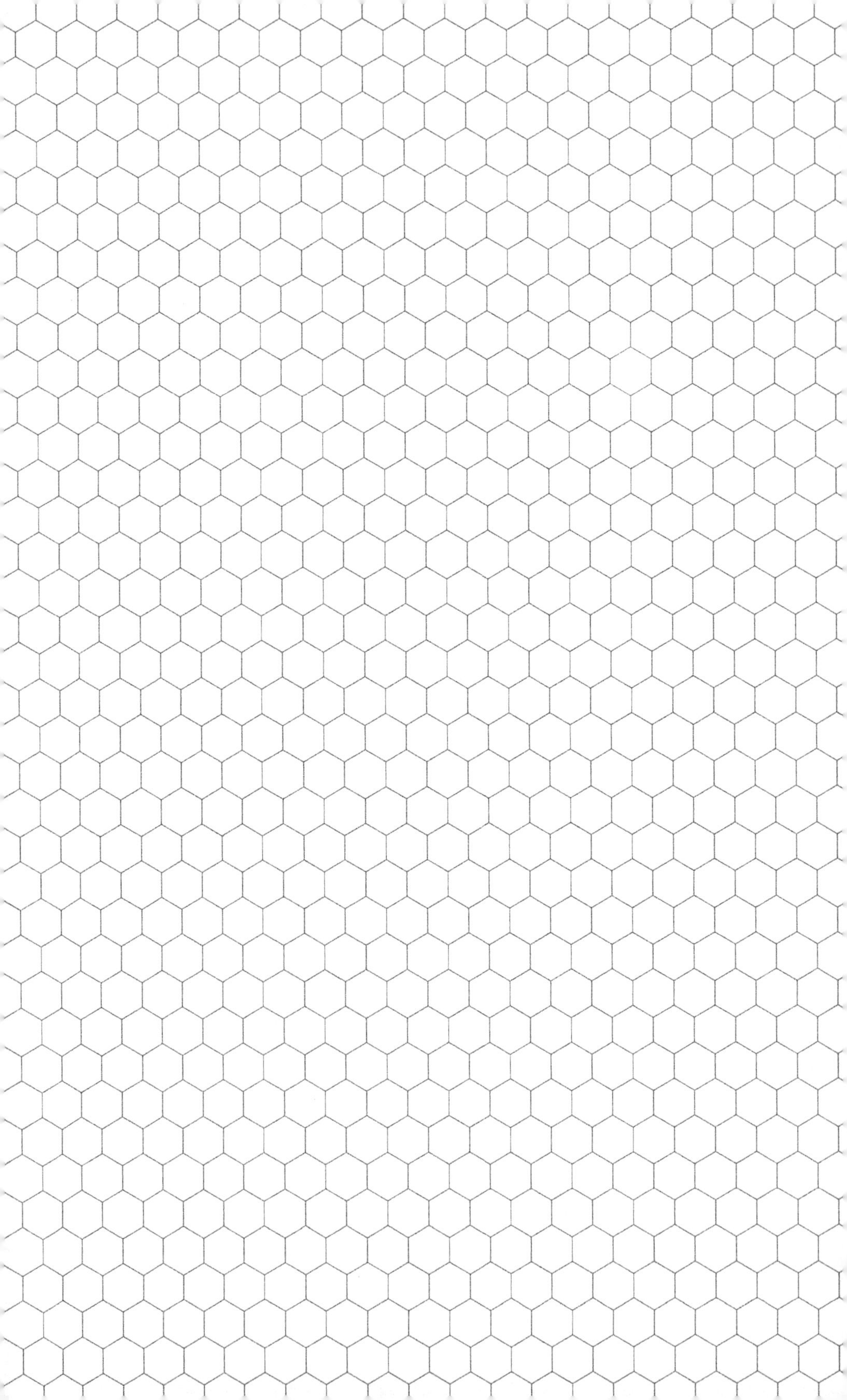

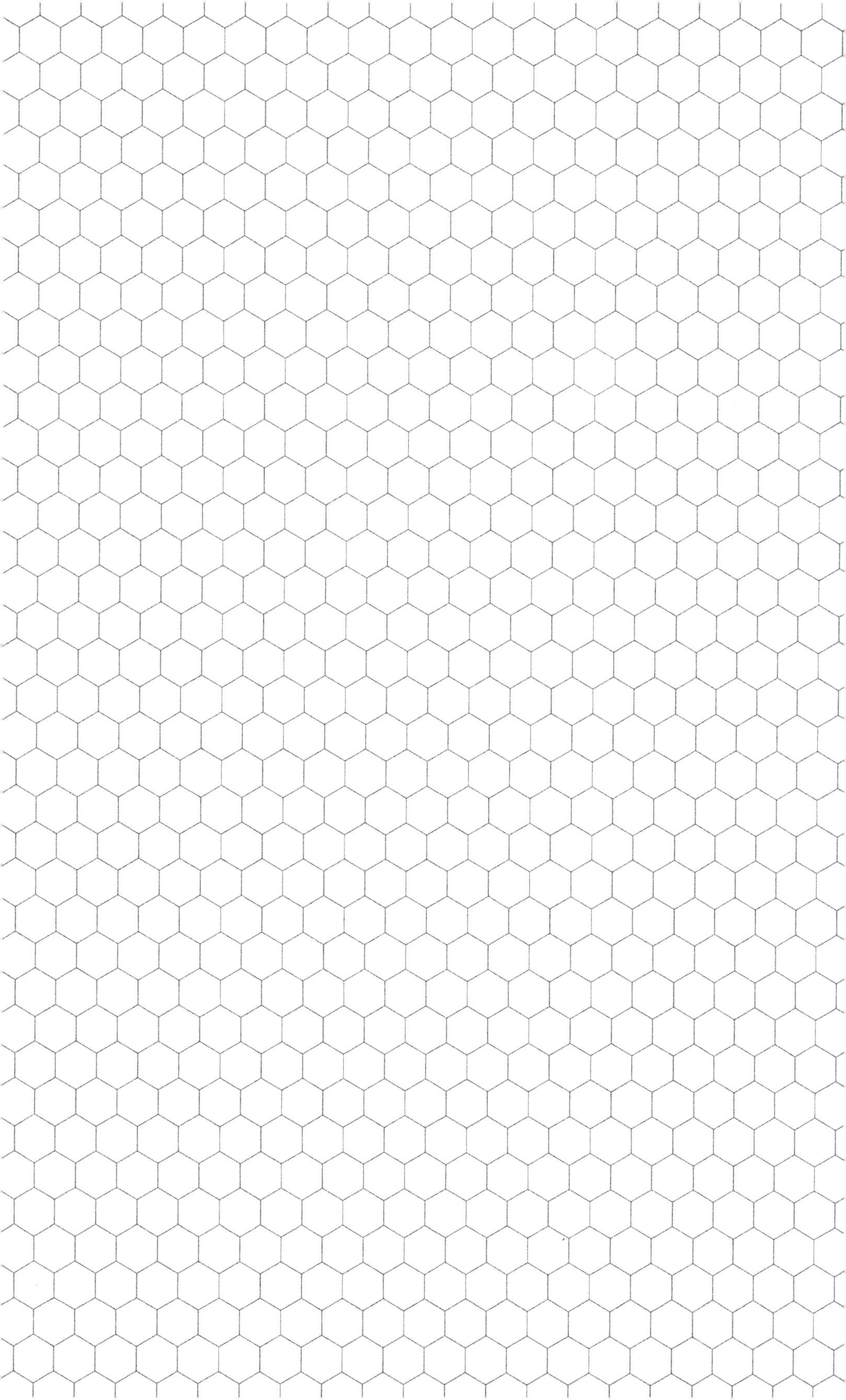

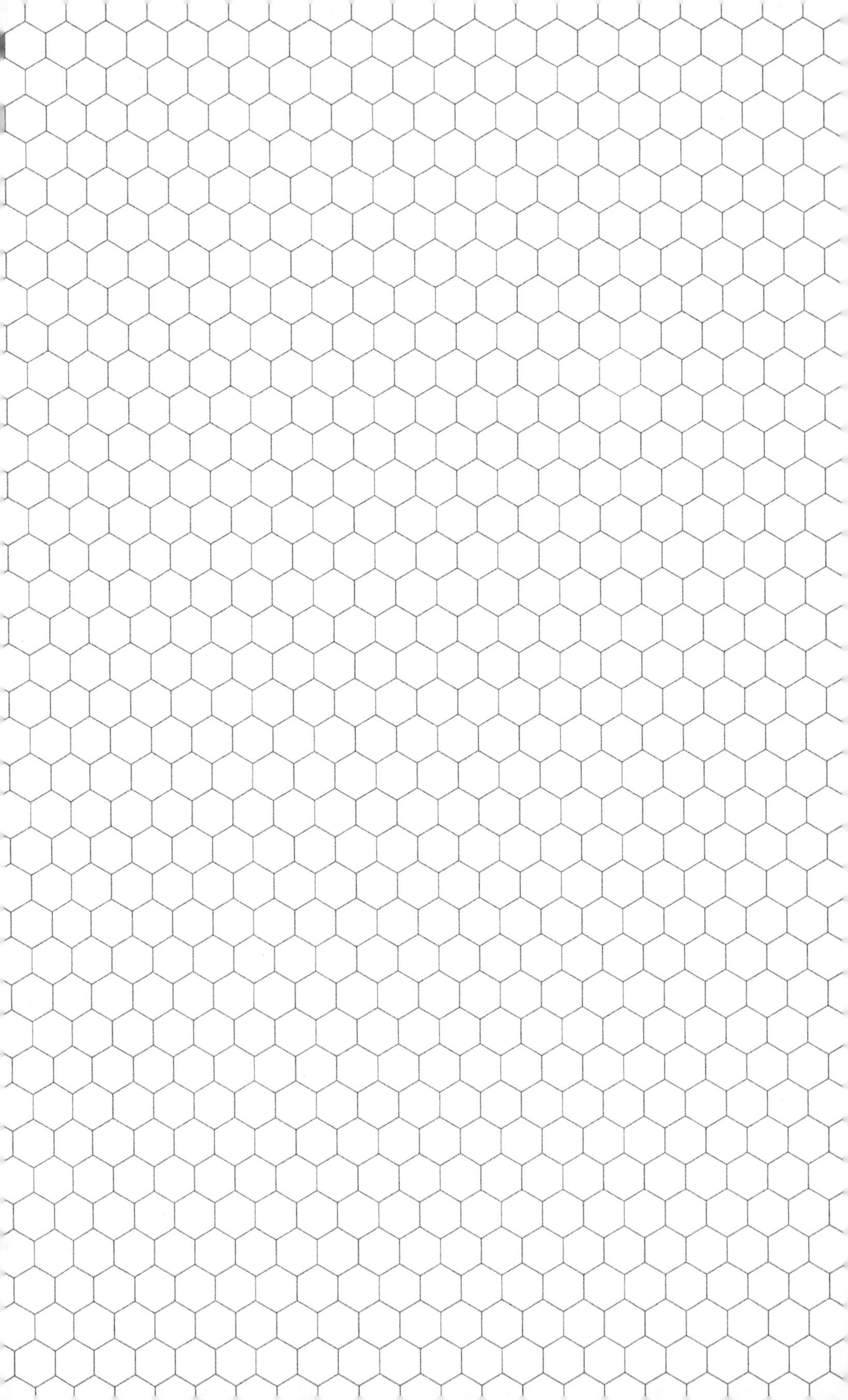

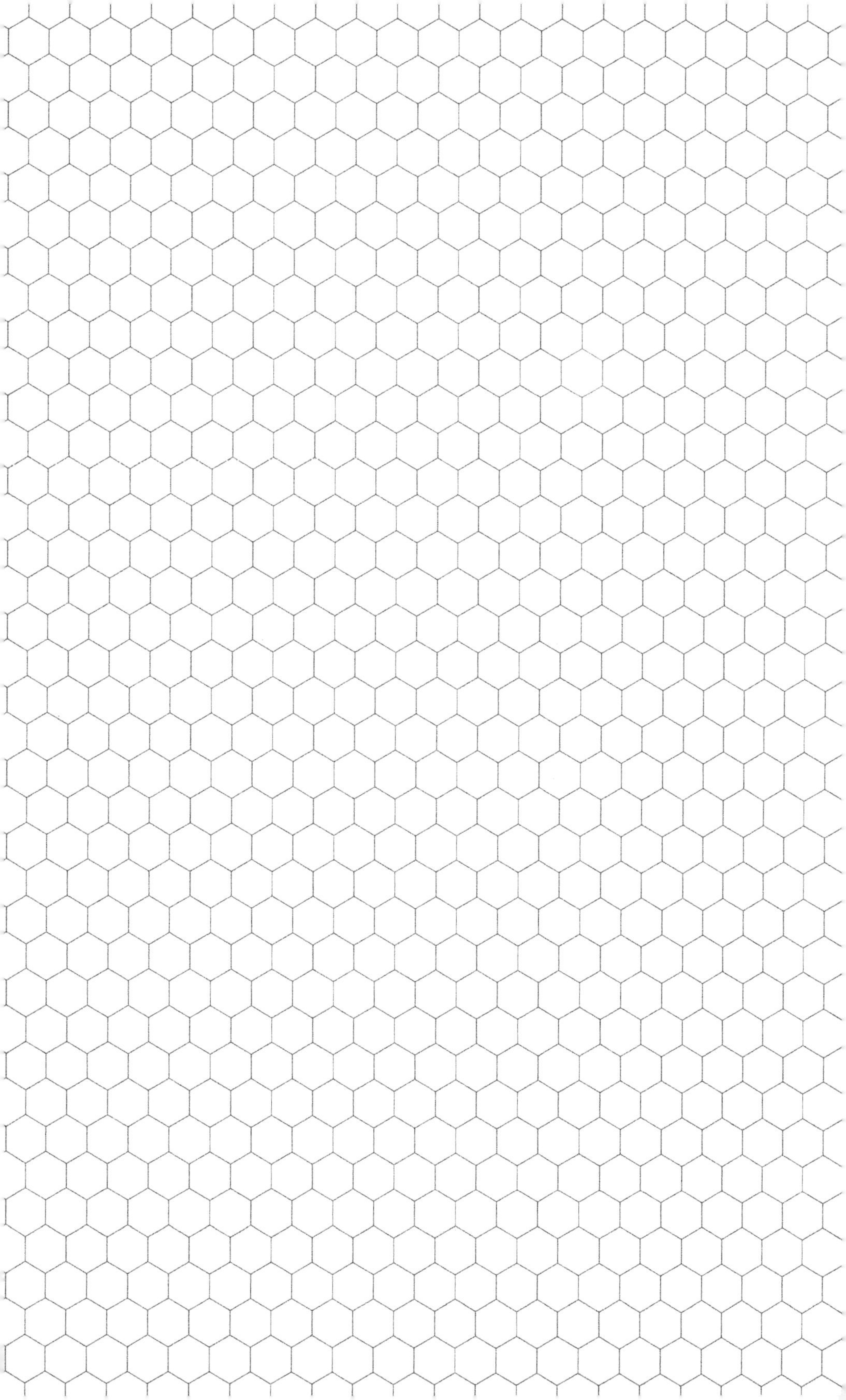

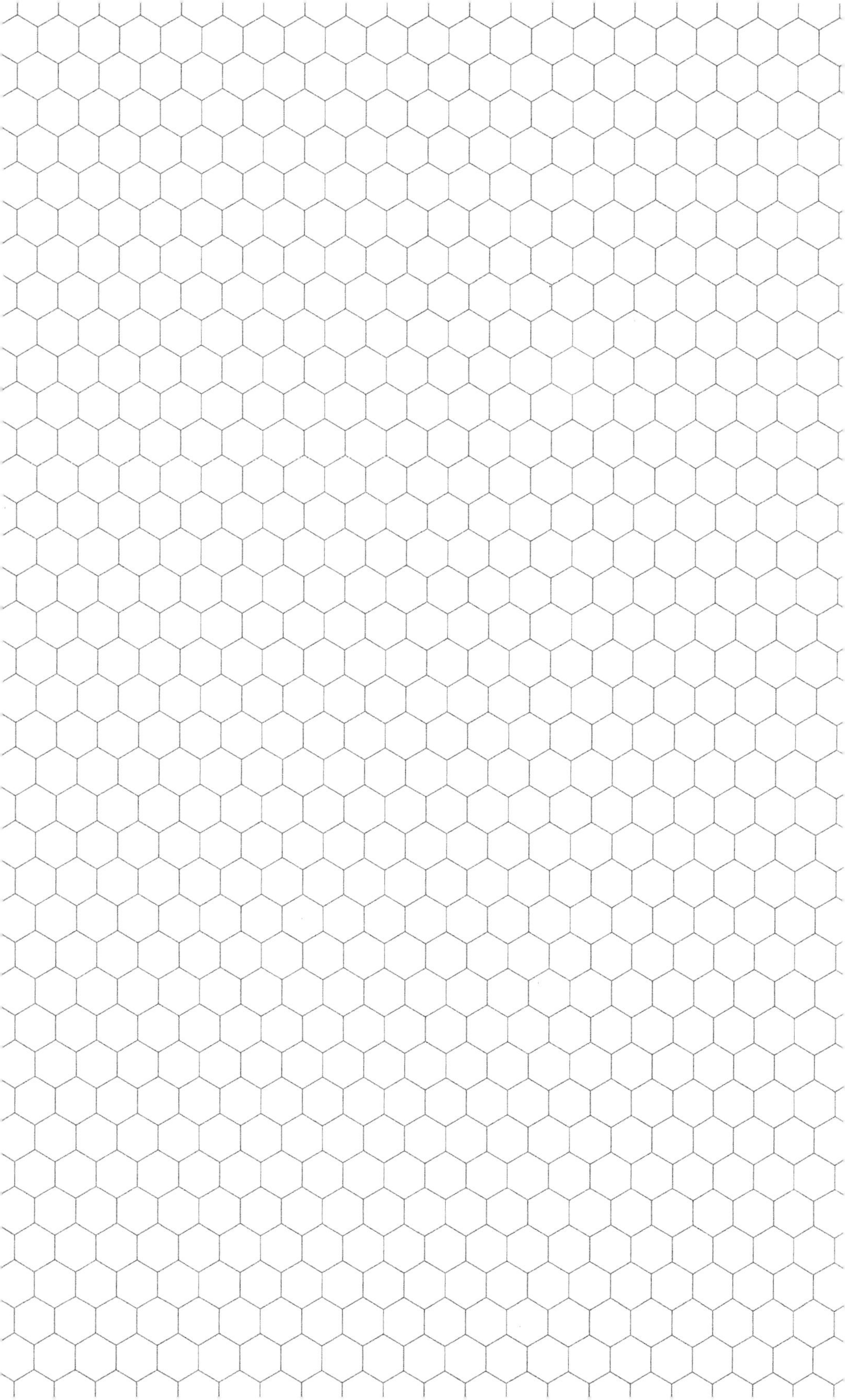

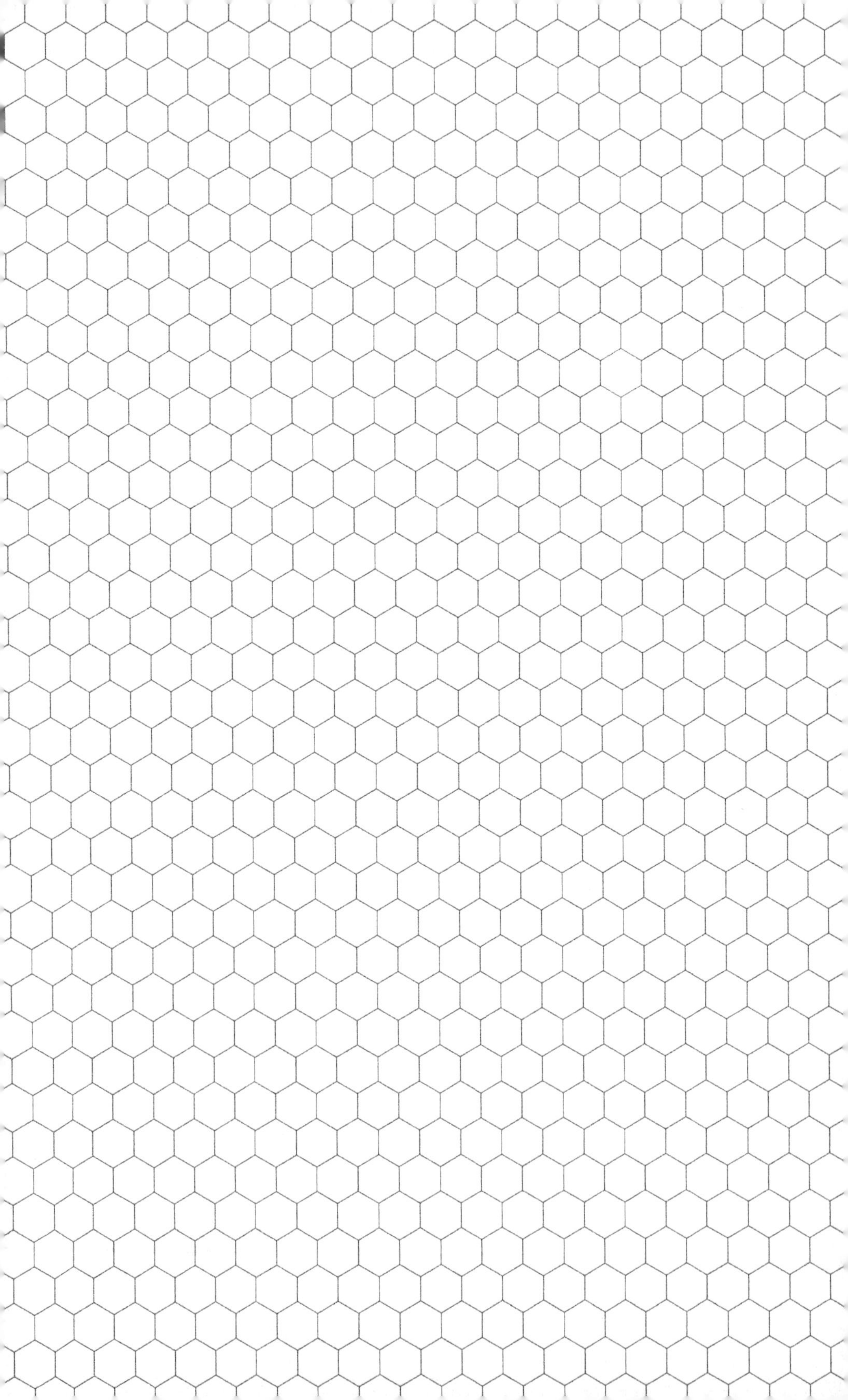

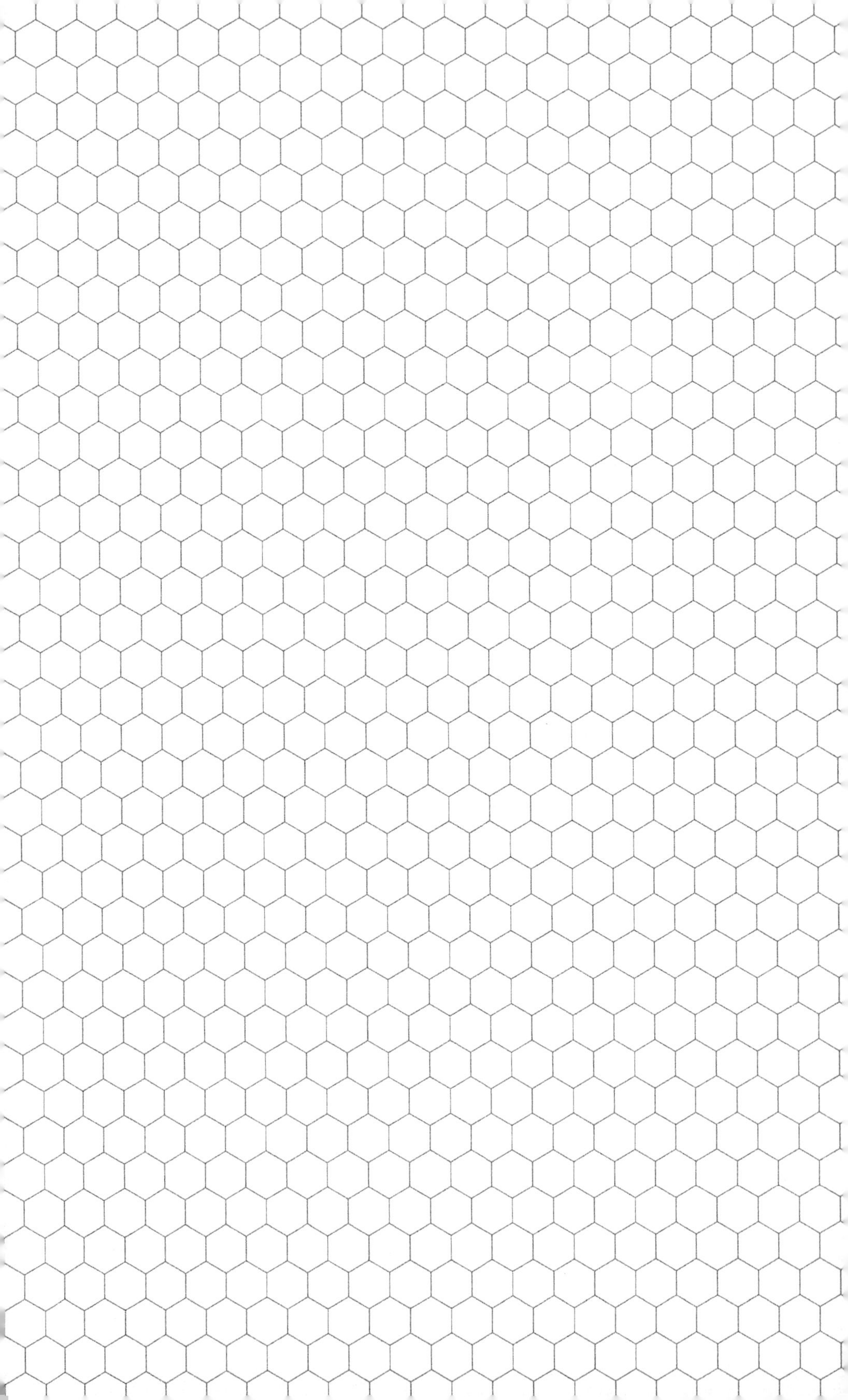

Made in the USA
Monee, IL
08 July 2026

56550996R00069